T0137437

A Survival Guide for Research Scientists

Ratna Tantra

A Survival Guide for Research Scientists

Ratna Tantra
Portsmouth, UK

ISBN 978-3-030-05434-2 ISBN 978-3-030-05435-9 (eBook)
https://doi.org/10.1007/978-3-030-05435-9

This Springer imprint is published by the registered company Springer Nature Switzerland AG
The registered company address is: Gewerbestrasse 11, 6330 Cham, Switzerland

To my mother, Partini Tantra
Without being aware, you have taught me all that I ever needed to know, so that I can survive... whatever the odds!

Preface

You will have noticed that I have dedicated this book to my mother. It was her gift to me that I should learn all that I needed to know on how to survive in the world, so that I can ultimately succeed. Likewise, dear readers, it is my gift to you that I should impart with my own knowledge, to help you survive as a research scientist in an increasingly demanding and competitive world. Nowadays, scientists are expected to do so many different things, and being a research scientist will mean the need for you to be more than just technically able.

Someone once asked me as to why I decided to refer to this book as a survival guide. Why just survive? Why not flourish/thrive/prosper instead? My answer to this is that, in order for you to flourish/thrive/prosper, you must first learn the basics on how to survive. However, I do not want you to only survive and survive miserably. I want you to survive as a research scientist and, at the same time, find self-fulfilment as a human being.

The purpose of this book is to teach you core skills not often taught at school or university. Hence, this book will be especially useful for those students who are about to embark on a Master's or PhD program in any scientific discipline.

When it came for me to identify what topics to cover, I am indebted to all those who came up to me and have asked questions in the past. Some of the questions were technical questions related to my discipline, but most were not such as "How do I ...":

- Design experiments?
- Ensure safe laboratory practice?
- Ensure best practice?
- Write reports?
- Prepare for a presentation?
- Publish peer review articles?
- Write bid proposals?
- Edit other people's work?
- Deal with difficult colleagues/office politics?
- Network?

- Look for a job?
- Manage my stress level?
- Deal with a bully at work?

By writing this book, I want to share knowledge and provide the answers to these questions (and more).

You can say that my approach is a holistic one, as I want to help you with your struggles not only as a scientist but that of a human being as well. Only by combining the two will you eventually find true happiness and fulfilment in your career as a scientist.

The book is divided into five main parts.

The first part gives you practical guidelines on how to take care of yourself, as well as your career.

The second part will give you guidelines on how to improve your technical research skills. I will cover topics such as how to conduct a literature review, design experiments, adopt best practice, ensure health and safety, etc.

The third part is focused on how to write and edit documents, with particular reference to the writing of reports, peer review publications, scientific bid proposals, etc.

The fourth part deals with challenges from the outside world. As such, I will cover topics on networking, how to deal with difficult people, how to deliver a presentation, how to behave in a meeting situation, etc. In addition, I will give you practical guidelines on how to lead a team and how to be a good project manager.

The final part will be dedicated on how you can secure your perfect job. I will show you how to develop a perfect CV, how to approach the job marketplace, how to prepare for interviews, etc. Furthermore, I will be covering additional topics on self-employment and redundancy.

Now let me wish you the very best of luck in your journey towards carving a successful science career, as well as finding self-fulfilment.

Finally, remember that life is a pilgrimage and that you are here to enjoy the journey!

Portsmouth, UK Ratna Tantra

Acknowledgements

Firstly, my immense gratitude to my husband, Keith Pratt, for his continuing love and support.

To family and friends, no words can ever repay your kindness and generosity.

I am indebted to those who either have been instrumental towards the production of this book or have dedicated their time to provide fruitful discussions on certain issues and topics. Special thanks to my Springer editors (Anita, Kiruthika and Faith) and Dr. Ian Cox (from JMP Division of SAS) on the discussion of experimental designs. To Professor Terry Wilkins (University of Leeds), you have not only been my mentor but also a good friend.

To Veronica Liana (Vero Designs) and Yosep Arizal (Vero Designs), many thanks for your artistic directions and contributions.

Finally, to the fathers! To my father I Wayan Tantra and Father John Humphreys (St. Colman's Church, Cosham), many thanks for showing me the light when things got a bit dark.

Contents

Part II In the Laboratory

Part III Writing

About the Author

Ratna Tantra has been a scientist for over 20 years. She has vast experience across different scientific disciplines, having worked in academia and government research. She got her PhD from University College London in electrochemistry. She then became a research associate specialising in microfluidics, first at Imperial College London and then at University of Glasgow. In 2002, she moved to the National Physical Laboratory (NPL), in which she worked in a biotechnology group before specialising in nanoanalysis. In 2016, after working at NPL for 14 years, she took redundancy and since then has become self-employed, with a portfolio of different activities, from launching a soap business (originating from her hobby) to working as a scientific expert for the European Commission. Her inherent passion for writing and genuine interest to help others has led her to write this book.

About the Author

Part I
Introduction

Chapter 1
Looking After Yourself

For you to find fulfilment as a research scientist, you must first survive as a human being. This is a basic necessity. Without this, you will find it hard to live your life. Hence, this first chapter is dedicated to you … not your career, not your science … just you.

When reading this chapter, you may argue: what is the point telling scientists on how to reduce stress and solve (potentially ongoing) personal problems? Isn't all of this highly superficial?

My answer to you is this …

If you read the preface of this book, I have made it clear that my concept of survival takes on a more holistic approach, in which my definition of what success looks like not only equates in achieving your dream career but to find fulfilment as a human being. As sad as it may seem, I sometimes encounter scientists who are quite high up in the career ladder, only to find them stressed out, not happy or even content with what they have. Why? Most of the time, they have forgotten to take care of themselves. Like martyrs, they have focused their energy on everything else, usually juggling their hectic scientific careers with a demanding family life, e.g. having to take care of children/elderly parents/an ill spouse. When taking care of everything/everyone else, they have neglected their own needs, as it is often easier to show compassion to others but not to themselves. Ultimately, the world around them shatters, as they experience unmanageable stress or work burnout.

Now, ask yourself this question: Is there much point of achieving career success if you are utterly miserable in your personal life? If the answer is YES, then read no further. If not, then do not skip this important chapter.

I will start this chapter off by telling you what I mean by stress and how to identify the symptoms. You have to be aware that everyone's stress response can be different and as such what methods you decide to adopt in order to effectively de-stress will also be different. I will be discussing some of the different options that are available to you. You should consider these, try them out and then develop a de-stressing strategy that is right for you.

R. Tantra, *A Survival Guide for Research Scientists*,
https://doi.org/10.1007/978-3-030-05435-9_1

1.1 What Is Stress?

It is difficult to accurately define stress. What is considered to be a stressful situation to some may not be for others. Furthermore, the word *stress* is often used rather loosely in everyday conversations (Muir, 2012). For example, you might say that you are stressed out, but in reality you may be extremely busy. However, the kind of stress that I will be dealing with in this chapter is one that can make you ill. It hinders you rather than propel you, e.g. to achieve your goals. The type of stress that I am talking about can render you helpless to its cruel mercy.

So, what is the origin of stress?

It is sometimes said that our stress response originates from our cavemen ancestry and the need to protect ourselves from imminent dangers (Bryant & Mabbutt, 2006). If you had lived as a caveman, then you would have all sorts of dangers lurking around, e.g. coming face to face with a tiger or mammoth. You would need to fight with other cavemen in order to protect your property or family. As such, human beings have inherited a fight-or-flight response, so that we know what to do when we come face to face with danger. Although the likelihood of any of us actually being confronted by a tiger nowadays is pretty slim, the fight-or-flight response is inherently still within us. It is useful in emergency-type situations, such as the need to step back when you see a speeding car approaching you on the streets at night.

When you experience a fight-or-flight response, your physiological equilibrium will change, resulting in you exhibiting one of the many symptoms associated with stress. To start, your sympathetic nervous system that regulates the body's unconscious actions (e.g., to keep your heart from beating without having to think about it) will be aroused. As depicted in Fig. 1.1, this state of arousal can result in

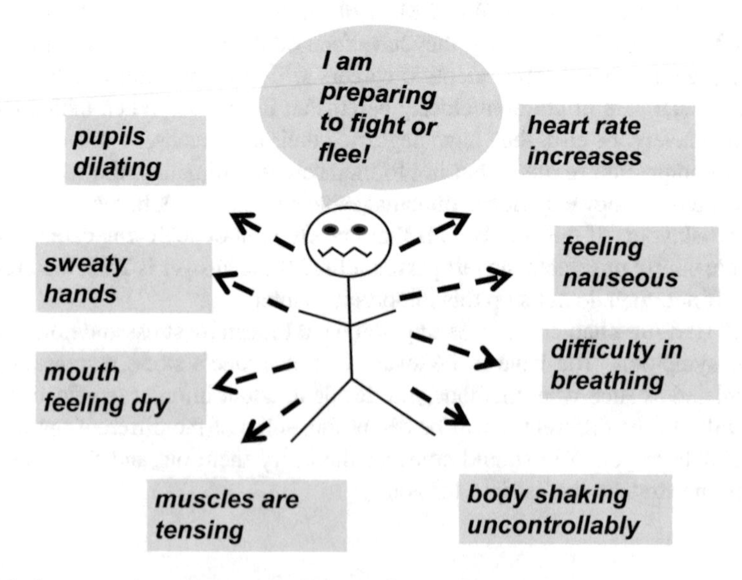

Fig. 1.1 Some symptoms that can arise from the fight-or-flight response

symptoms of anxiety, such as an increase in heart rate, an increase in blood sugar level, sweaty hands/feet and dilation of the pupils. Without being aware, the amount of stress hormones, such as adrenaline and cortisol, will automatically increase (Bryant & Mabbutt, 2006).

Although the fight-or-flight response is needed in an emergency-type situation, it can sometimes show up in situations when there is no real danger, i.e. when there is only perceived danger. Let me illustrate this point further. Let's say that you have to do an oral presentation in front of a large audience as part of your science project. It is your first time standing in front of an audience and you feel nervous. Your hands shake, your mouth goes dry and your voice quavers uncontrollably. Undoubtedly, your fight-or-flight response has been stimulated. Yet, if you look at the situation carefully, you are not in any real danger. Your life will not solely be dependent on how good or badly you do in your presentation. Once you have done your presentation, your fight-or-flight response deactivates and then everything returns back to normal. Obviously you will not die from the experience, as the human body can cope with some level of stress. In fact, it may be argued that responding to stress can improve your presentation performance, as you are likely to be more prepared. As such, stress can motivate you to some extent, thus propelling you to perform your very best in challenging situations. Furthermore, stress can keep you busy. It prevents your mind from wandering, straying away from unwanted, negative thoughts. So, undoubtedly there is such a thing as *good stress* (McGonigal, 2015). However, the real danger comes when your stress level is not managed properly, resulting in negative consequences, ultimately impacting your health. It can affect you physically, so much so that you can be experiencing daily symptoms of nausea, shallow breathing, aches and pains, etc. Daily stress can take a toll on your immune system too, making you more prone to the common cold. It can affect your mental health, in which you may suffer from unwanted daily symptoms of anxiety, depression or panic attacks. Hence, if you are not managing your stress level well, you can be counterproductive at work. For example, stress can reduce your ability to make the right decisions or worse still cause accidents. If you are unable to manage your stress level, it can escalate, as illustrated in Fig. 1.2. The illustration shows how the initial symptoms of stress can feed into itself and magnify, resulting in a cyclic state.

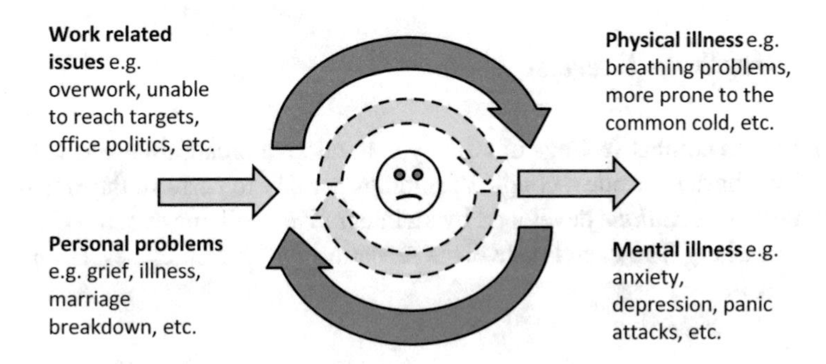

Fig. 1.2 How stress can easily magnify and escalate

For example, having breathing problems due to an initial response to stress can make you feel even more anxious, particularly if you have not linked your initial symptoms to stress but instead have attributed it to some kind of a physical illness, e.g. heart issues.

1.2 How to Prevent and Combat Stress

You need to accept that life is sometimes not easy and thus will be full of ups and downs. When the going gets tough, it is inevitable that you will be stressed out. However, this does not mean that you cannot minimise its negative impact. You will need to find ways to de-stress.

So, how can you de-stress?

The first step is to be aware of how your body reacts and responds to stress (Davis, Eshelman, & McKay, 2008). This means the need for you to identify the symptoms that are unique to your own stress response (see Fig. 1.1). Once you are able to do this, you will then be able to put a coping strategy in place. There are a number of self-help books that can help you to do develop your own strategy. You can easily spend weeks reading the different literature. Having trawled through the literature myself, I found several tips that are worthy for your consideration.

1.2.1 Exercise

Find time in your daily routine to exercise (Johnstone, 2015). 30 min of intense activities (enough to get your heart pumping above its normal rate) is your target. At least, this is what has been recommended to me by my own doctor. Remember that this does not necessarily mean that you will need to join a gym in order to achieve your daily target. You can easily fit this into your routine schedule, e.g. jogging/cycling to work, doing house work and gardening.

1.2.2 Breathing Exercise

A great way to combat feelings of anxiety is to adopt a breathing exercise regime. Personally, I have set aside a couple of minutes per day to perform the well-*known 4-7-8 breathing technique* developed by an integrative medicine expert, Dr. Andrew Weil (Weil, 2011). The exercise involves repeating four breath cycles. Each breath cycle involves:

- Inhaling (through the nose), on a mental count of 4 s
- Holding the breath on a count of 7 s
- Exhaling (through the mouth) on a count of 8 s

When you are doing a breath cycle, you will need to put your tongue on the ridge that is between the front two teeth of your mouth. For a better description on how you should perform this breathing exercise, please refer to the video on YouTube ("Asleep in 60 seconds: 4-7-8 breathing technique claims to help you nod off in just a minute - YouTube", n.d.). How often should you do this? If this is your first time, then you should only do it two times a day. You may increase its frequency after you feel more comfortable and have mastered the technique (usually after a month or two). Personally, I found this breathing technique to be an effective way to relax, useful if you need to deal with difficult people or before delivering an oral presentation. It somehow deactivates the fight-or-flight mechanism when you are under stress.

1.2.3 Visualisation

This is when you use your imagination to conjure up positive healing images, to help you relax (Davis et al., 2008). You can think back in time when you have felt total relaxation. Hence, your chosen image can be one of you being on holiday and having a massage under exotic palm trees, whilst the hot sun warms your skin. Obviously, the image does not need to be related to past events. You can also sketch an imaginary scene in your mind, e.g. an image of you being on a boat, floating away on a clear blue sea without a care in the world.

1.2.4 Eat Well

You will need to take care of your diet and eat well (Davis et al., 2008). Generally, reduce your intake of sugar, fat, alcohol and caffeine. Consume more healthy food such as fruits, vegetables, wholegrain and fish. I am not saying that you should stop enjoying treats such as wine, sugary cocktails, cakes and puddings, but the trick is to have these things in moderation and definitely not everyday.

1.2.5 Sleep Well

Make sure that you get enough sleep; otherwise you might incur what is known as a sleep debt (Hancock & Szalma, 2008). Having a sleep debt can result in serious consequences, potentially causing you to have an accident. Adopt a good sleep hygiene practice, by having a winding-down session at least an hour before you sleep, e.g. taking a bath, listening to soothing music and reading something that will relax you. Avoid caffeine in the late afternoon or evening. Make sure that a couple of hours before bedtime, you do not: exercise, watch TV or drink alcoholic beverages. Improve your sleeping environment, such as the use of comfortable bedding, blocking out any distracting lights (e.g. using an eye mask or heavy curtains) and blocking out noise (e.g. using earplugs or having double glazing).

1.2.6 Find Inner Peace

There are different things that you can do in order to achieve this. Doing mindfulness activity or meditating is just an example. You may also like to connect yourself to the Higher Self, such as having a relationship with God or following a religion, not only to find peace but also to improve your spiritual intelligence (Johnstone, 2015).

1.2.7 Achieve Work-Life Balance

This is about your ability to manage time to juggle work and leisure activities (Lockett, 2008). Consider taking up a hobby, spending time with the family, eating out in a restaurant, having a massage (such as aromatherapy, shiatsu, reflexology), doing yoga or Tai Chi, visiting museums, etc. (Nandram & Borden, 2010). In addition to leisure activities, you can also make behavioural changes when you are at work in order to help you achieve a better work-life balance. Try to adopt a good working attitude, such as being more assertive and speaking softly (rather than being aggressive) when you communicate your point across. You can also change your own attitude on how to solve issues. For example, instead of getting angry with colleagues or customers who annoy you, you may want to learn to compromise, especially when others do not see your point of view. Remember to take regular breaks during your working day, e.g. to eat your lunch and drink water regularly. Try to stick to your working hours and ensure that you leave work on time. When you are home, learn to switch off from work so that you can mentally unwind, and then get ready for bed and sleep so that you can recharge your batteries (Cropley, 2015).

1.2.8 Adopt a Paradigm Shift on How You View Life Events and Remain Positive

What do I mean by adopting by a paradigm shift?

Often it is not the event itself that is upsetting but how you decide to view it. Think about it. There are so many people that have endured hardships in their lives. Yet, they seem to have made the best of their situation and move on. Some have not only managed to survive their ordeal but also miraculously turned a negative situation into a positive outcome, e.g. Nelson Mandela and Helen Keller. So, what makes these people survive their ordeal and be resilient, whereas most of us would have given up or ended up ill as a result? Primarily, it is in their ability to accept their situation early on and view their predicament differently. As such, they have adopted a paradigm shift on how they think about the upsetting event. They remain positive throughout, always concentrating on what they can do or what they can create rather than focusing on what they cannot achieve.

The concept of changing how you think about life's events is actually the basis of cognitive behavioural therapy (CBT) (Branch & Willson, 2010). The CBT concept focuses on the idea that thoughts, feelings and behaviour are interlinked. CBT is about challenging your thoughts, beliefs and thinking patterns, in such a way that you can change on how you perceive a particular problem. By changing how you think, you will automatically change how you feel and subsequently affect how you behave. Although I will not be covering CBT in great details, it may be worthwhile for you to read up on the topic from past references (Branch & Willson, 2010; Sweet, 2012). If you are in need of further guidance, then I suggest that you seek professional help in the form of a qualified CBT therapist.

1.2.9 Realise When You Are Underwater and Seek Help

When life finally proves to be too much for you and you are struggling to cope, particularly if you are not coping emotionally or mentally, then do not suffer in silence and seek professional help immediately such as in the form of a medical doctor, counsellor, psychologist or psychiatrist. It's best to do this earlier on, as you may be able to nip the problem in the bud. Remember that feelings of anxiety and depression can potentially escalate (as depicted in Fig. 1.2). If you think that you may have a mental illness, then it is best to familiarise yourself with the range of treatments or therapies available. By gaining knowledge, you can further discuss the matter with a professional and identify on how best to improve your mental health (Freeman, 2017; Ivory, 2012; Rosenthal & Hollis, 1994).

In a summary, please refer to Fig. 1.3 for further considerations. Remember, in developing your de-stressing strategy, you will need to find a combination that works for you. As such, it may take you some time, of several trial and errors, in order for you to come up with a suitable coping strategy that is right for you. If you have any doubt on what you can or cannot do, e.g. whether or not you are physically fit to perform 30 min of daily exercise, then it is vital that you discuss the situation with a relevant professional, e.g. a medical doctor.

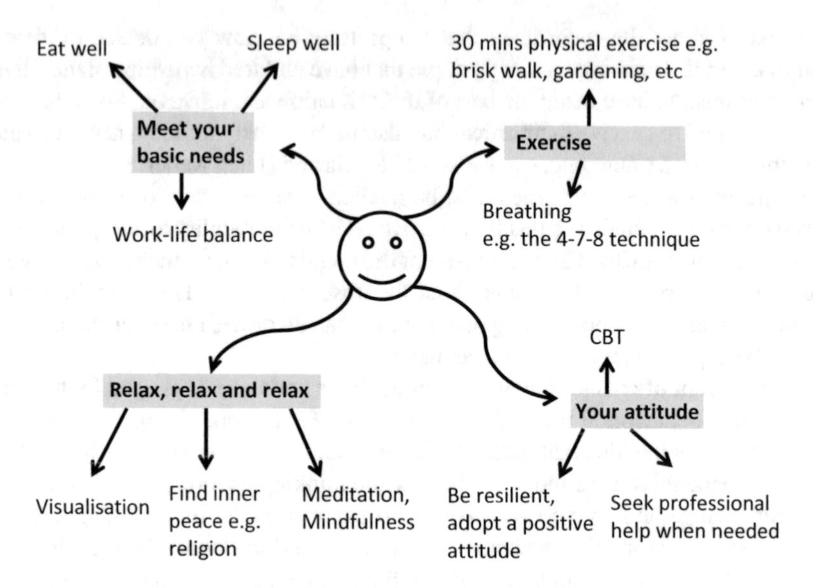

Fig. 1.3 Potential ways to combat stress

1.3 PRAAM (Problem, Research, Assess, Action, Monitor): A Logical Approach to Tackle Personal Problems

A key skill to have in life is the ability for you to solve personal problems. I understand that personal problems can inherently be very complex, in which there are no easy answers. But whenever you feel pessimistic about a particular situation, you must remain positive and leverage on your natural talent as a researcher. Remember that you have the ability to research, think analytically and form rational decisions. As such, I will be discussing the use of PRAAM (Problem, Research, Assess, Action, Monitor), a logical approach that you can use to solve personal problems:

1. **P**roblem: This is when you need to identify and further define the problem.
2. **R**esearch: This is when you will need to dig deep for information, to understand the problem better and find potential solutions. Besides simply googling for answers, you can rely on other sources: books, seeking professional advice, talking to friends/family, etc.
3. **A**ssess: This is when you need to evaluate your research findings, particularly identifying the factors (i.e. contributors) that are most influential towards your current problem. If it helps, you can score the contributors relative to one another.
4. **A**ction: This is when you need to come up with suitable action points, so you have a logical plan. You will also need to factor in any past knowledge and your own belief system/values when coming up with action points.

5. **M**onitor: This is when you need to pinpoint what the ideal outcome should be and how you are going to measure to monitor success. Remember, there is not much point in doing something if you are not heading in the right direction.

Note that any one of the above stages can be iterative, in which you may need to go through a particular stage more than once. For example, if you see no benefits in carrying out a particular action, then you may need to revisit a particular stage in PRAAM e.g. going back to stage 1 in order to improve your understanding of the problem.

In order to illustrate the PRAAM approach better, please refer to the two case studies below. These case studies have been drawn up from my experiences on how I have approached certain personal problems in my life.

Case Study 1: Nocturnal Hypnic Jerks

Problem

At some point in my life, I had suffered from nocturnal hypnic jerks. This is when I would get woken up abruptly every time I was dozing off to sleep. My symptoms varied, ranging from hearing a loud bang in my head to certain limbs jerking uncontrollably. I ended up having sleep deprivation, often sleeping for 2–4 h per night. This was not only detrimental to my general health but also had a direct impact to my working life, as I would find it hard to concentrate during the day.

Research

When I conducted my research, there was little information on the Internet. There was some vague explanation as to why I may be experiencing these hypnic jerks, which included from having too much coffee to over-exercising. As a result, I avoided caffeine and I made sure that I was not too overtired before going to sleep. However, my initial attempt to solve this problem had not worked. I realised that I had to dig deeper and over time the research phase became an iterative process, in which I needed to further define and understand the underlying problem better.

Assess

It became clear that I needed to seek expert advice for my condition. I needed to see a neurologist. The research also highlighted the importance of getting access to some kind of medication that would (in the short term) allow me to have a good night sleep. I needed medications that were sedating, not addictive, and thus were adamant to stay away from the benzos!

Action Points

My first port of call was to talk to my general practitioner (GP) and urged him to make me an appointment to see a neurologist. Whilst it took some time for me to get seen by the neurologist, I did asked the GP for a non-addictive medication that would help me sleep better. The GP had prescribed amitriptyline, which worked wonders. When I eventually got the appointment to see the neurologist, she explains to me that my jerks were the result of accumulating stress. As a result, my brain was sending inappropriate signals whenever I tried to sleep. She recommended that I continued with the amitriptyline—slowly weaning off as the jerks become less frequent. From the consultation, it became obvious that I had no major illness and that I needed to reduce my stress level in order to eventually get rid of the jerks. As such, I reduced my caffeinel intake, found peace in the form of spiritual fulfilment, exercised regularly and sought the advice from a professional therapist.

Monitor Success

Throughout my ordeal, I plotted graphs in Excel to monitor my success. The measurement indicators chosen were the number of hours I slept, the number of tablets that I was consuming and the number of hypnic jerk episodes I have per week. I realised from my visit to the neurologist that the hypnic jerks were not going away overnight but by monitoring my success I felt more confident and at ease. It took me close to a year to finally get rid of my hypnic jerks and was glad when it was finally over. My sleep now has never been better (without the need to take any medications).

Case Study 2: Weight Problems

Problem

Like most people, I have always struggled with weight, unfortunately with a tendency to being overweight rather than underweight. Key to my weight loss journey was to understand the cause behind weight gain.

Research

Like most people, I thought that weight gain was governed simply by the concept of calories in versus calories out. As a result, I opted to reduce my calorie intake by adopting a low-fat diet. Although this initially worked, in that I had lost 5 kg within a 1-month period, I had noticed that going further beyond the 5 kg point was close to impossible. As such I had to do more research, particularly to redefine the problem. I was surprised to learn that not only weight gain is governed by the amount of calories that you consume but also the process is affected by certain hormones. In particular, the level of insulin in your blood has a dominating effect on weight gain (Fung, 2016).

Assess

The findings clearly point out that if you want to reduce weight, you will need to control not only the net calories per day but also your ability to control your insulin level (Fung, 2016). Remember that insulin gets released whenever we eat food, with some foods being worse than others, particularly when consuming simple carbohydrates like sugars.

Action Points

After assessing the different types of diets that would allow me to not only limit my calorie intake AND control my insulin (i.e. blood sugar level), I decided to opt for intermittent fasting (Fung, 2016). For me it is both practical and easy to follow. There is no strict calorie counting and the diet does not restrict any food groups. I definitely did not want to go towards the route of no/low-carb dieting, as a diet mainly protein and fat will never satisfy my appetite. However, I have to admit that I am more careful with my carb intakes nowadays.

Monitor Success

Throughout my weight loss program, I plotted graphs in Excel in order to monitor my success. The measurement indicators chosen included my weight, body mass index and body waist circumference. The ability to monitor had meant that I was able to manage my weight better.

1.4 Summary

Accept that as part of life, there will be times when you will be prone to stressful events. If not managed well, then the effect of stress can escalate. It can get out of control and make you counterproductive, potentially impacting your physical and mental state of being. This chapter presents the different ways on how you can de-stress and how to deal with personal problems, so that you can have some control over your life. Although I have given you several practical tips on how you can achieve a more stress free life, ultimately it will be up to you to develop a routine that works for you.

References

Asleep in 60 seconds: 4-7-8 breathing technique claims to help you nod off in just a minute - YouTube. (n.d.). Retrieved January 7, 2018, from https://www.youtube.com/watch?v=gz4G31LGyog

Branch, R., & Willson, R. (2010). *Cognitive behavioural therapy for dummies*. Hoboken: John Wiley & Sons.

Bryant, M., & Mabbutt, P. (2006). *Hypnotherapy for dummies*. Hoboken: John Wiley & Sons, Ltd.

Cropley, M. (2015). *The off-switch: leave work on time, relax your mind but still get more done*. London: Virgin Books.

Davis, M., Eshelman, E. R., & McKay, M. (2008). *The relaxation & stress reduction workbook*. Oakland: New Harbinger Publications.

Freeman, M. (2017). *The mind workout*. UK: Hachette.

Fung, J. (2016). *The obesity code: unlocking the secrets of weight loss*. Vancouver: Greystone Books.

Hancock, P. A., & Szalma, J. L. (2008). *Performance under stress*. Farnham: Ashgate.

Ivory, B. (2012). *Healthy living tips for improving physical and mental health good nutrition, weight-loss tips and fitness program, meditation, stress reduction, and mental health well-being*. Pittsburgh: Dorrance Publishing.

Johnstone, M. (2015). *The little book of resilience : how to bounce back from adversity and lead a fulfilling life*. UK: Hachette.
Lockett, K. (2008). *Work/life balance for dummies*. Hoboken: John Wiley& Sons.
McGonigal, K. (2015). *The upside of stress : why stress is good for you, and how to get good at it*. London: Vermilion.
Muir, A. J. (2012). *Beat stress*. London: Teach Yourself.
Nandram, S. S., & Borden, M. E. (2010). *Spirituality and business : exploring possibilities for a new management paradigm*. Berlin: Springer.
Rosenthal, H., & Hollis, J. W. (1994). *Help yourself to positive mental health*. Taylor & Francis, Milton Park.
Sweet, C. (2012). *Change your life with CBT: How cognitive behavioural therapy can transform your life*. UK: Pearson.
Weil, A. (2011). *Spontaneous happiness*. Boston: Little, Brown and Co.

Chapter 2
Looking After Your Career

Chapter 1 is all about meeting your basic needs. This chapter is all about your *wants*, particularly your *career wants*. From my own experience, I can tell you that searching for a job is tougher nowadays, compared to 20 years ago when I first began to look for full-time employment. As such, I would like to give a special message for our younger readers. It may be that you are still enjoying your freedom and carefree lifestyle, as a student. But trust me, when it comes to your career, you are never too early to think about what you want to do in life.

In this chapter, I will be discussing several items:

1. How to identify what you want from a career, thus identifying your dream job.
2. How to plan in order to achieve your dream job.

An important aspect in achieving any dream job is to be self-aware and to conduct yourself at the highest level of ethics and integrity at all times. Unfortunately, as scientists we rarely get the opportunity to be taught about moral and ethics. In this chapter, I aim to shed light on the matter. I hope that this will propel you to adhere to an acceptable level of work ethics in all that you do, such as acting more responsibly when in the laboratory.

2.1 Discovering the Real You and What You Want

If you are in a position in which you know what you want to do in the next 5–10-year time, then you can more or less skip this section and go straight to the next section. However, if you are one of many, in that you either have no idea in what you want, then you must keep on reading as I will tell you how to identify your dream job.

R. Tantra, *A Survival Guide for Research Scientists*,
https://doi.org/10.1007/978-3-030-05435-9_2

An important first step in identifying your dream job is to discover the real you, specifically what matters to you, what you want out of a job and life (Storr, 2014). As such, you will need to:

- Identify where your values lie. This will enable you to focus on what things are most important to you, thus helping you to define your career choices. It may be that you will want to put family first, in which case your ideal career is one that will allow you to have time to raise your family. It is important to be aware that your values can change through time. For example, you may value security and hence a permanent contract in your 20s. By the time you are in your 40s, your values may shift, e.g. to strive for a better quality of living. For example, you may want to live in an unpolluted, quieter environment and so ultimately it is the location of your next job that will be the determining factor.
- Identify what motivates you. What gets you up in the morning? Here lies your passion. When you are motivated and passionate in what you do, you will become more proactive to perform your very best.
- Identify your aspirations. Identify the challenges that you are craving for in your career. Make a *bucket list* and identify those stretching goals that you will want to achieve.
- Know your personality type and how you prefer to behave. For example, are you an extrovert or introvert? Do you need to work with others or are you a hermit, thus happy working on your own? Are you one of those people who naturally want to care and nurture others?
- Identify your strengths and weaknesses. It's not enough to just have passion; you must be realistic and know where your abilities lie. In addition to your technical skills, you will need to be clear on what soft skills or transferrable skills you have. For example, are you a natural networker, e.g. do you often act as a hub between people? Do you find it easy to influence and motivate others?

For some of the questions listed above, it can be difficult to make accurate judgements about yourself without being subjective. That is why, if possible, you should adopt a more methodical/objective approach, by carrying out relevant tests. For example, if you want to know more about your intellectual strength, then you can take an IQ test (Russell & Carter, 1989). If you want to know how good you are with numbers, then you will need to do a numeracy test. If you want to understand your personality type, then you can complete a personality-based questionnaire. A good test to do is the Birkman test, which aims to link your personality type to potential careers (Birkman Fink & Capparell, 2013).

Although it is generally a good idea to do these standard tests, you will need to interpret the final results with caution. As human beings we are extremely complex and the interpretation from such tests may not be 100% accurate. Nonetheless, these tests will give you some guidance, i.e. a ballpark figure on which side of the fence you sit on. For example, it is rare to see a smart person failing an IQ test.

2.2 How to Identify Your Dream Job

In identifying your dream job, you must collate the information about yourself and what you want out of life. In addition, you will need to identify the reason why you want a job in the first place. Remember: You don't set your alarm clock every night so that you can get up at 7 am (or earlier) just for the sake of it. You go to work because you have attached some kind of meaning to your job (Krznaric, 2012; Hirsh & Jackson, 2012), which can include the need to:

1. Have a healthy salary (along with job security)
2. Have a recognised status
3. Have a certain level of dignity in society, e.g. in which you are able to take care of your family or loved ones
4. Make a difference to other people's lives
5. Follow your passion
6. Use your talents in a job that will stretch you, so that you can realise your full potential

The first three points above can be classed as extrinsic drivers (Krznaric, 2012), in which you view your work as nothing more than a means to an end. The last three are classified as intrinsic drivers (Krznaric, 2012), in which the work that you do are valued, as an end it itself. In other words, this is when you are living to work rather than the other way around. It is important to note that what a job means to you can change over time. For example, in the beginning of your career, you are likely to focus on the extrinsic factors such as having a regular income. As you get older and more financially secure, you may want a job that can give you fulfilment, e.g. in which you may want to make a difference to the world.

In an ideal world, your dream job will tick all of your boxes. In reality such jobs may not exist. As such, you will need to have a flexible attitude, balancing what you want out of a job versus the choices that you have (Fig. 2.1). Hence, in drawing up your final list on the kind of job that you want, you will need to do a bit of market research. There is little point to wish for something, if it does not exist or if such a job gets advertised only once in a blue moon. You can see if your dream job exists in the marketplace by frequently visiting online job advert sites ("Jobs | Job Search | Job Vacancies on jobs.ac.uk", n.d.; "Job Search | Job Site & Career Advice | Monster.co.uk", n.d.).

In deducing if a job is right for you, you will need to read both the job and person specifications carefully, to establish if the job will make you happy. Do you see yourself doing the job on a full-time basis? Will you be able to handle the intellectual and emotional aspects of the job? If your dream job does not exist or rarely exists, then you will need to investigate alternative career paths. If you have access to a career centre, it is worthwhile discussing your options with a career advisor.

You must also start to think about the kind of organisation that you will want to join: academia or industry. Often, the main focus in industry is on applied science and what will generate profit. If you want a more relaxed environment that allows

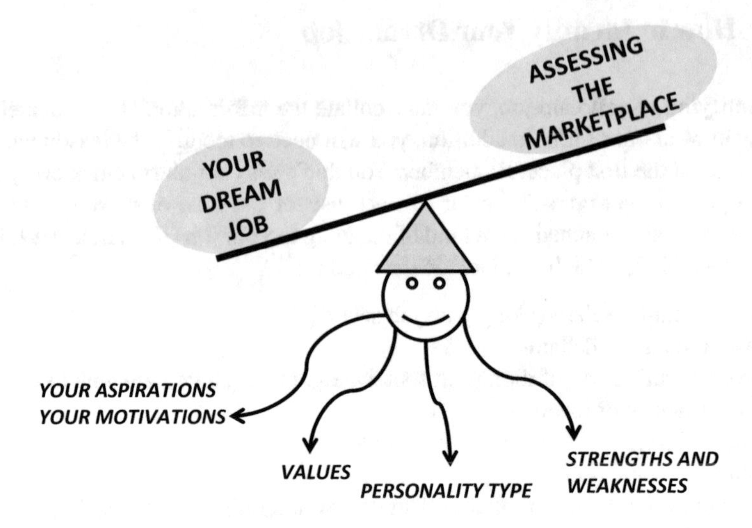

Fig. 2.1 Identifying your dream job vs. assessing the marketplace

you to freely explore innovation, then consider going to academia. Remember that the working culture between industry and academia is different. In industry, people tend to work their set hours, clocking in and out each day. There is also less flexibility in what you can do with regard to research, as activities have to be justified in order to support the business model and generate income. Unlike academia, industry tends to prize patents more, as opposed to peer-review publications. Although the work culture in academia is generally more relaxed, you may find yourself having to work for longer hours and at weekends in order to chase laboratory results. The key drive in academia is to generate high-impact-factor publications, as academic research is often at the forefront of innovation and glamour. Of course, you have other organisations that sit between industry and academia, e.g. government research.

Although the model of different working cultures that I have just presented to you is rather simplistic, my aim here is give you rough guidelines on what you can possibly expect from different type of organisations. However, the best way to get to know the culture of an organisation is to work for them. Therefore, whilst you are still a student, you should try to experience various organisational cultures whenever opportunities arise. Consider taking on an internship or going for an industrial placement as part of your course. You can even do an unpaid summer job, just so that you can experience different organisational cultures and learn more about yourself and where you naturally fit in. Understanding where you fit is important, as it is unwise to work for an organisation in which there is a mismatch. You will only find it hard to adapt and change once you are hired to do a job in that organisation.

Finally, remember that you can always opt out of science. This does not mean that you have wasted your time in getting your degree. You will have acquired many valuable transferrable skills that will be useful to other sectors. For example, if you are a theoretical scientist, then you may have computer modelling skills, which will be in high demand in lucrative sectors such as the IT or the banking industry.

2.3 Identifying Goals to Support Your Dream Job

So, once you have identified your dream job (and possibly the ideal organisational environment), you will need to develop a strategy on how you can realise your dream.

In a book (written by Professor Peters) called *The Chimp Paradox* (Peters, 2011), an important first step in order to realise any dream is to identify the goals that support it. Peters (Peters, 2011) had underlined the need for you to differentiate between *dreams* and *goals*. He defines a dream as something that you may wish to happen in the future but in which there is no guarantee that you will get it, as it will not be under your full control. Goals on the other hand are different. You will have more control in realising them and will only be set because of your ability to achieve them in the first place.

As a personal exercise, I now want you to write down your own dream job and then identify the goals that support it. When writing your goals, remember the acronym SMART, as you will need to write SMART goals (O'Neill, Conzemius, Commodore, & Pulsfus, 2006). SMART stands for specific, measurable, attainable, realistic and timely. By measurable, I mean that you will need to have some way of measuring your success, once you have completed a specific goal. By timely, I mean that the goals must adhere within a certain time frame and as such you will need to set an appropriate deadline for each. For the purpose of illustration, please refer to the case study below.

Dream Job: To Become a Lecturer in a Top-Ranking University
Identified goals supporting my dream:

1. By January 5th 2017—identify an exciting and upcoming science area, where there is much scope for high-impact publications.
2. By June 10th 2017—start a PhD with a prestigious and internationally well-renowned group, with a supervisor who is passionate about publishing in high-impact journals.
3. By April 6th 2020—take on extra duties that will support an application to become a lecturer. Look for opportunities to: mentor undergraduates, prepare lessons, teach classes, be a laboratory demonstrator, mark student's exam papers, etc.
4. By June 10th 2020—publish at least five peer-review papers in high-impact journals (impact factor of five or more); collaborate with colleagues and external collaborators in order to achieve this.

5. 6 months before completing PhD studies—read job adverts for lecture posts and develop your CV. Talk to a career advisor for advice and support.
6. 3 months before completing PhD—start to send out CVs to prospective employers.
7. … etc.

Remember that in defining the goals to support your dream job, you will need to refer to relevant job adverts to see what requirements you will need. Make a list of all the hard and soft skills (as well as any work experience) that you will need for the kind of post that you have in mind. If you are missing any vital skill or experience, then this is the time to acquire them and filter them into your list.

After accomplishing any goal, you must remember to celebrate your success, even partial success. According to the *Chimp Paradox* (Peters, 2011), this has something to do with keeping your inner chimp happy, so that you can remain positive and stay motivated.

2.4 Be Flexible in Your Approach

Although you may have a dream job in mind, it is important to remember that nothing in life is certain. There will be times when you will face failure.

So, what can you do in the adversity of failed attempts?

Whether things are going your way or not, always remember to move forward in life. Moving forward means that you will need to be kinder to yourself and take good care of your health/wellbeing. Whatever you do, be sympathetic to your situation, as you cannot change what has happened. Instead, concentrate on the future and see what other options are presented to you. If this means completely scraping your original idea and going back to the drawing board to redefine your dream, then so be it. The key is to be flexible on your approach to every problem that comes your way. Being flexible was something that I had to do after my PhD, when the prospect of having a job in industry became bleak. Allow me to tell you my story, in order to illustrate this point.

I did a PhD on the development of glucose sensors. I took on this PhD because of my interest in the subject and the possibility of working in industry to develop new sensors for the market. The demand for glucose sensors was huge … millions of patients. At the end of my PhD in 1997, I had expected to be working in industry as a research scientist, working to develop new sensors. It was something that was set in my mind as to what I would be doing for the rest of my life and I was looking forward to living this new chapter. In reality, I became frustrated as I realised that there was little demand for my background. For a start, there were only a few companies in the UK that did glucose sensing. In addition, I felt that these companies were looking for people who have had experience in product development. In the end, I only got one interview with a glucose sensing company and the job was given to a candidate more suited for the job, i.e. he had more industrial experience. Although initially disappointed, I had to bounce back quickly and be in touch with reality. I realised that if I wanted to stay in research, I needed to go into a new field. I needed to look for a job that was able to appreciate my electrochemistry background but was less strict on the actual scientific discipline. After relying on my network and touching base with a few ex-colleagues, I became aware of a new group starting at Imperial College and was informed that they were looking for an electrochemist to join the group, in order to develop an electrochemical based sensor incorporating lab-on-chip technology. At that time, lab-on-chip technology was only beginning to take off and I knew that I had a good chance in getting the job. I contacted the head of the group as a result. Fast forward a couple of weeks later, I was thrilled to have landed the postdoctoral position, although it meant that I had to abandon my dream of working in industry. Quite surprisingly, I enjoyed the job. I learnt several new skills and gained work experience in the area of fabrication technology. Reflecting back on this event of my life, I can say that not only this was the time where I grew most as a scientist but also the decision to embark on something new became a real turning point for my career. The experience gave me greater confidence in myself as a scientist and prompted me to achieve more, in the years to come.

So, my point is this: no matter how much you plan, it is possible that your plan, hopes and dreams may get crushed at the last minute. The above case study also shows you on how important it is for you to touch base with your network, to be sensitive to what is going on around you, in order to identify other avenues. The key is to always adopt a flexible approach to life. Most likely you will need to compromise so that you can adapt to your environment (rather than the other way round). If you want to be more cautious with your future, it might be worthwhile to think up of multiple dream jobs in parallel and to identify the relevant goals associated with each one (as depicted in Fig. 2.2).

2.5 Foster Strong Work Ethics

I am a firm believer that whatever you sow, you will subsequently reap. Reflecting on my father's Balinese philosophy on life, he has always reminded me that when you plant potatoes you will get potatoes. Hence, in order for you to reap the benefits

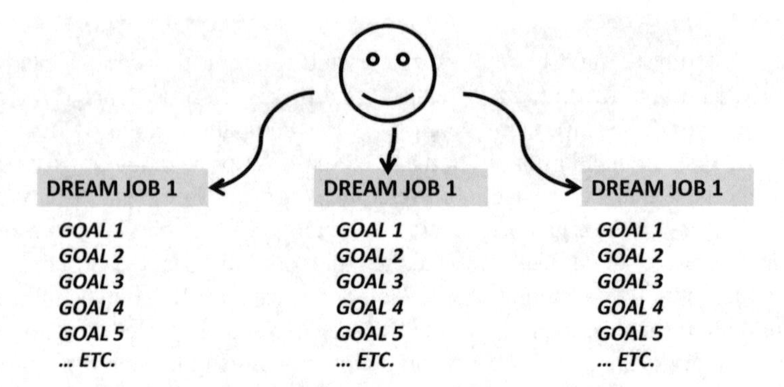

Fig. 2.2 Being cautious in your approach in getting your dream job: have multiple dreams and identify the relevant goals that support them

and find fulfilment as a scientist, you must sow good things to begin with. Remember that whatever you do in order to reach your goals and dreams, always remember to maintain your integrity and foster strong work ethics. As such, you will need to know the difference between right and wrong and my advice is to always choose the right path.

With certain professions, work ethics are clear. Doctors for example have to take the Hippocratic Oath before taking up employment. The oath states (amongst other things) the need to either help or not harm your patient (Miles, 2005).

But what about research scientists? What are our moral and ethical obligations?

Well, this is less clear and can very much depend on the scientific discipline. For example, if you are a toxicologist, then you are likely to deal with animal experimentations. As such you will need to consider the ethical aspects on how to treat the animals. In addition to ethical rules specific to your own scientific disciplines, there are also generic rules, in which ALL scientists should adhere. Resnik (1998) previously pointed out to some of the basic dos and don'ts:

1. Do not commit scientific fraud such as fabricating or misrepresenting data.
2. Respect copyright.
3. Be careful to avoid careless errors in your work.
4. Know that you have the right to pursue new ideas and criticise old ones.
5. Practice openness whenever possible, e.g. share your methods/results and declare any conflict of interest.
6. Do not plagiarise, i.e. take the work/idea of someone else and pass it off as your own.
7. Exercise moral responsibility to society and the environment, e.g. when conducting experiment in a laboratory, consider how your activities can impact the health and safety of your colleagues, the public and the environment.

So why are ethics and moral obligations important in science?

If you do not follow a level of ethical standards, then your research can result in a negative outcome. Let me refer you to a past example to illustrate, which relates to a premature announcement of the results from a cold fusion research. The research was conducted in the 1980s by Martin Fleischmann and Stanley Pons (Resnik, 1998), in which they had reported some preliminary laboratory results far too early. They claimed that excess heat had been produced using a simple electrochemical cell set-up and heavy water. The result from such an announcement caused a huge media sensation, as such a discovery would have meant the answer to all of our energy needs. However, following their announcement, many laboratories tried to repeat the findings without much success. Their results were either inconclusive or contradictory. In fact, most cold fusion scientists much later believed that the research by Fleischmann and Pons was based on careless errors and sloppiness. Undoubtedly, their reputation in the scientific community was tarnished.

2.6 Summary

This chapter is all about you and how to look after your career, specifically how to identify your dream job and realise it. In order to identify your dream job, you will need to know who you are and what you want out of a job. A key step in realising your dream job is to establish if such a job exists. If it does not, then you need to be flexible in your approach and compromise on what you have to offer versus to what is actually needed in the marketplace. Once you have identified a dream job that exists, you will need to identify the suitable goals to support it. Whatever you need to do in order to achieve your goals and ultimately your dream job, remember to know the difference between right and wrong. Always choose the right path; strive to be ethical and be morally responsible in all that you do.

References

Birkman Fink, S., & Capparell, S. (2013). *The Birkman method: Your personality at work*. Hoboken: John Wiley & Sons.

Hirsh, W., & Jackson, C. (2012). *Planning your career in a week*. UK: Hachette.

Job Search | Job Site & Career Advice | Monster.co.uk. (n.d.). Retrieved October 9, 2017, from https://www.monster.co.uk/.

Jobs | Job Search | Job Vacancies on jobs.ac.uk. (n.d.). Retrieved October 9, 2017, from http://www.jobs.ac.uk/.

Krznaric, R. (2012). *How to find fulfilling work*. London: Pan Macmillan.

Miles, S. H. (2005). *The Hippocratic oath and the ethics of medicine*. Oxford: Oxford University Press.

O'Neill, J., Conzemius, A., Commodore, C., & Pulsfus, C. (2006). *The power of SMART goals: Using goals to improve student learning*. Bloomington: Solution Tree Press.

Peters, S. (2011). *The chimp paradox: The mind management program to help you achieve success, confidence, and happiness*. London: Vermilion.

Resnik, D. B. (1998). *The ethics of science: An introduction*. Abingdon: Routledge.
Russell, K., & Carter, P. (1989). *The IQ test book*. London: Sphere Books.
Storr, P. (2014). *Get that job in 7 simple steps*. UK: HarperCollins.

Part II
In the Laboratory

Chapter 3
Literature Review

During my undergraduate study, I had to conduct a literature review for a third-year project. It was to discuss the industrial synthesis of pyridine and pyridine derivatives. I was a typical undergraduate at the time. Without having any inkling on how best to do this, I had not thought about the approach and headed for the library to search for every topic on pyridine. I then wrote my review based on any published papers that came my way. When the papers did arrive, I read them one by one, diligently starting from the title and then making my way to the conclusion. During the writing phase, I wasted a lot of time, trying to write and rewrite. Most of the time I had lost my thread, as to why I was reading the papers in the first place. In the end, my literature review research project lacked the needed focus and clarity and so I got a C grade.

In years to come, my skills to conduct a literature review had improved considerably. I was able to be more focused, extract essential information easily from papers and not waste time.

It is my aim in this chapter to teach you on how best to conduct a literature review. As such, I will be presenting five basic steps on how you can do this effectively. My hope is that through time, you will not only improve and master the process of reading and extracting information from a sea of literature, but also be able to improve the speed in which you can accomplish this.

3.1 Step 1: Write Down Your Statement of Purpose

So, why do a literature review?

Although the need to do a literature review as part of an undergraduate course is a good enough reason, there are many other reasons, such as the need to (O'Leary, 2005):

R. Tantra, *A Survival Guide for Research Scientists*,
https://doi.org/10.1007/978-3-030-05435-9_3

- Assess past methods, to help you plan for your own experiment. This is useful so that you are aware of current methods and *thus not reinventing the wheel.* You can also learn from historic information to improve experimental design, e.g. by identifying potential errors and variables that may affect the outputs of your own experiments.
- Find an explanation for results that you have observed in the laboratory.
- Compare results with others in the field, thus forming a basis for discussion to explain experimental findings.
- Identify future work/ideas.
- Provide an up-to-date understanding of the topic in your field, which can be useful in many instances, e.g. to provide the context for proposed research activities in a bid proposal.
- Find relevant references in order to write a paper for peer-review publication.
- Identify any past debates or scientific consensus on certain issues.
- Identify important terms and terminology used.
- Appreciate the style of writing relevant to your scientific discipline, such as how things are described. The idea here is NOT to plagiarise but to help you develop your own style of writing and conform to what your audience expects of you.

Before beginning your literature review, it is vital that you are clear as to what you are trying to achieve. As such, you will need to write a statement of purpose, to explain why you are conducting the literature review in the first place. By doing so, you will be defining the scope of your review. The statement of purpose should be based on questions that you will want answering. These questions should not only be specific but also be of significance. Having a statement of purpose is useful, as it allows you to remain focused and is a constant reminder on what is important. For example, you can refer to the statement of purpose to help you decide if you should purchase a particular document or not.

3.2 Step 2: Identify Keywords and Then Search for the Relevant Literature Using Appropriate Databases and Search Engines

From your statement of purpose, you will need to identify keywords. This is important so that you can type them into various search engines and databases in the Internet (Fink, 2005), of which there are many to choose from (such as Google, Yahoo Search and Live Search (MSN)).

Although using a search engine can be useful, you will need to remember that anyone can publish material on the Internet. As such, there is no guarantee that the information is credible. In contrast to search engines, citation databases are more advanced, particularly on how information is organised and the material kept in these databases would had gone through a formal editorial process (College Library Services CSN, n.d.). As a result, this somewhat improves the credibility of the published information. Bear in mind that some databases are not free. For example, some are provided by vendors (as part of a subscription service) through a library. The best known databases in academic circles are the Web of Science from Thomas Reuters (ISI) and Scopus (as introduced by Elsevier Science in 2004) (Aghaei Chadegani et al., 2013). They are both very good and share many similarities. However, these also have their differences, e.g. in terms of costs associated to access (Aghaei Chadegani et al., 2013). Having said this, not all bibliographic databases incur a cost. There are those that are free to access but are often associated with particular scientific disciplines. PubMed for example, launched by the US National Library of Medicine (Khare, Leaman, & Lu, 2014), is a citation database that focuses specifically on life sciences and biomedical information. The overall choice of one database over another will depend on a number of different factors, e.g. personal preference, what is commonly used by a specific scientific community and whether or not your library currently provides a particular service.

In addition to bibliographic databases, researchers should also make use of patent-based database. According to the European Commission, a patent search is important in research because it (European Patent Office & European Commission, 2007):

- Can give you background information on your research topic.
- Can give you an appreciation of some of the issues associated with your discipline.
- Can give you up-to-date information, e.g. defining technological advances and state of the art.
- Can find potential solutions to technical problems.
- Avoids duplication of research efforts.

One of the most common patent databases used in science is esp.@cenet ("Espacenet - Home page", n.d.). This is offered by the European Patent Office and is free to access. There are others that you may also like to consider, such as Patentscope (run by WIPO) ("WIPO - Search International and National Patent Collections", n.d.) and Google Patents ("Google Patents", n.d.).

Whatever search engines or databases you have decided to use in order to do your search, it is important to be thorough. Remember to:

- Refer to the statement of purpose from time to time, so you do not lose your focus.
- Record your list of keywords in conjunction with the different search engines and databases used.
- Be flexible in what keywords you choose and update them if needed.

At the end of Step 2, you should have a filtered list of the literature that you will want to potentially access. You can then shortlist this further by carefully reading the associated "abstract" information from each paper, to deduce if it matches with what you need, once again as defined by your statement of purpose.

3.3 Step 3: Order Your Papers—Manage Your Search

At this point, you should have a shortlist of references that you would like to get. Purchasing papers can be done via your institute's library or directly from online sources, e.g. via the journal's website. However, there is a drive nowadays for the scientific community to publish articles not only online but also through open access, which means that readers may have free access to certain articles.

When the papers finally come through, you will need to manage your literature findings. You will need to keep a record of every paper, noting down the title, author/s, year of publication and journal. To make your life easier, I urge you to keep records via a reference manager software. Such a software will allow you to import data and then cite any references with ease. There are a number of different software to choose from, with the most common being EndNote, RefWorks, Reference Manager, Zotero and Mendeley. Both Zotero and Mendeley are free to access.

When you have finished reading your papers, you will need to file these in some way, so that you can refer to them at a later date, if needed. Use whatever system you like as long as you can then access them from a well-organised filing system. If the paper comes in a digitised format, then you will need to create topic folders in your computer for filing. Getting a digitised form of a document is always recommended, as it will be a lot easier to store, e.g. saves on space.

3.4 Step 4: How to Read and Extract Information from Documents Effectively

So, how do you go about reading papers effectively, so that you can extract the information that you need?

It may be worthwhile to remember that this often depends on the type of document that you have to read, as some documents have a predefined way of how information is organised. This is particularly the case for research scientific papers and patents, which will now be discussed further.

For a scientific paper, it is best to develop an overview of the paper first, which can be done by:

1. Reading the title
2. Then the abstract, particularly focusing on the conclusion drawn from the abstract (Durbin, 2009)
3. And then reading the conclusion itself

By reading these three sections sequentially, you should be able to understand what the paper is about, and at which point you will then need to decide if you have further interest or not. If you are still unsure, then read the last few paragraphs of the

Introduction section as it should detail what the paper is trying to do. After this, you will need to decide what to read next and how much time you will want to devote. This will ultimately depend on how familiar you are with the topic and what your interests are, e.g. as dictated from your statement of purpose. If you want to skim and extract information, then you need to focus on the appropriate section of the paper. For example, if you are interested in the final results, then you can jump straight into the figures and tables; make sure that you read the accompanying figure legends so that you can identify the variables of interest. Without reading the text, you can try to analyse and draw your own conclusions just from the figures and tables. Once you have done this, you can refer to the accompanying text (in the results and discussion section) for further information. If you are a newcomer to the field, and you want to identify knowledge gaps, then the introduction section will be of interest to you.

Patents are organised in a different way to research papers. You generally have three sections associated with patents: front page, specifications and claims. The front page often contains the title, abstract (or summary), patent number, date, inventor and application/or assignee (company or individual applying for that patent). The specifications part describes what the invention is and includes background of the technology and drawings (Kramer, 2005). The claims section is considered to be the legal boundary of that patent.

So what is the best way of reading patents? In general:

1. Always read the title and then abstract first, to decide if a patent will be of an interest to you.
2. Read the specifications, i.e. describing the summary and background of the invention.
3. Look at the figures.
4. The final step is to read the claims that describe what the patent protects.

Once again, be flexible in your approach, as it will all depend on your personal circumstances and statement of purpose. For example, if you are an engineer in a company and you want to check if your product/process does not infringe a third-party patent, then it is likely that you will be focusing on reading: the title, then abstract followed by the claims (Kramer, 2005).

3.5 Step 5: Be Active to Effectively Absorb the Written Information

Quite often, you can be overwhelmed by the amount of material that you will have to read. If you are in a position in which you are likely to experience information overload, then you will need to be more proactive when reading. There are several practical tips that you can do in order to absorb the written information better

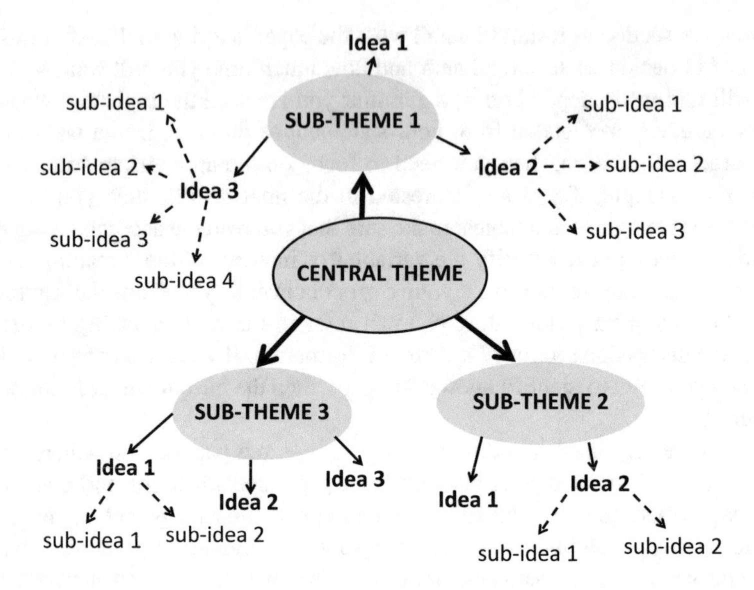

Fig. 3.1 Example template of a spider diagram

(especially if like me, you have a poor memory and a short concentration span). For example, you can:

- Highlight main points that stick out or of interest to you (to be governed by your statement of purpose).
- Make side comments and notes as you read along.
- Do mind mapping, such as condensing information into a spider diagram.

A spider diagram is a wonderful way to summarise information, giving you a visual presentation of the text (as illustrated in Figure 3.1). As an example template to illustrate. As you can see, you start a spider diagram by drawing a circle at the centre of a page, which represents the general theme of the topic. After this, other circles that radiate from the central circle are drawn, to represent sub-themes. Again, each sub-theme can radiate into specific ideas and sub-ideas and so on … and so forth (Shah, 1998). Creating a spider diagram is a good way to help you digest how the different topics and sub-topics relate with one another. You can be as creative as you like when producing spider diagrams, for example by using different colours to represent different sub-themes, sub-concepts or sub-ideas. The diagram can be simple with the central topic connected to a few circles, or it can be complex, in which there are lots of different interconnections.

Creating a spider diagram will enable you to digest information better as it can give you a helicopter view of what you have read so far. It can provide clarity to enable you to come up with new ideas easily. To illustrate this point further, please refer to an example depicted in Fig. 3.2. In this example, a spider diagram is drawn to help identify the main variables that are responsible for governing the

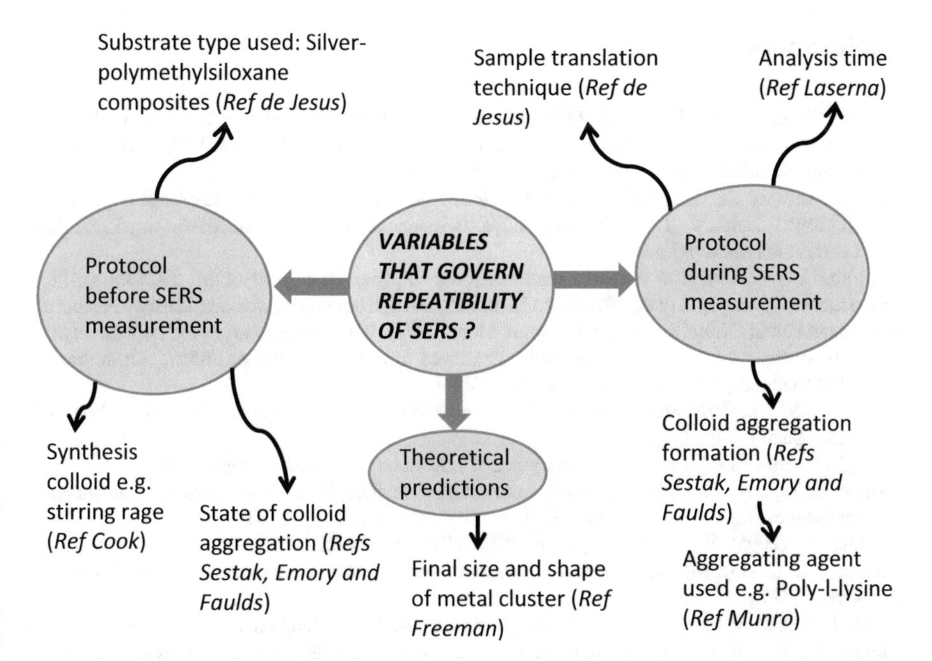

Fig. 3.2 Illustration of a spider diagram to identify the variables that govern the repeatability of SERS

repeatability of an analytical technique called surface-enhanced Raman spectroscopy (SERS) (Tantra, Brown, Milton, & Gohil, 2008). The spider diagram here clearly shows the governing variables of interest and how they interconnect. The spider diagram makes it easy for us to see that the poor repeatability of the SERS technique:

- Is associated with the protocols, both before and during the SERS measurement.
- Can be explained by the fundamentals that correlate the size and shape of the final metal clusters to the overall SERS signal enhancement factor.

3.6 Summary

Doing a literature review is an integral part of your research. Yet, if not done properly, this can be a daunting process. One hurdle lies in your ability to extract the kind of information that you want from a sea of literature. Another lies in your ability to absorb and make sense of the given information. However, there is a strategy that you can follow to make the process more effective. In this chapter, I have presented a five-step process, to detail on how you can carry out literature review with greater ease and confidence.

References

Aghaei Chadegani, A., Salehi, H., Md Yunus, M. M., Farhadi, H., Fooladi, M., Farhadi, M., et al. (2013). A comparison between two main academic literature collections: Web of science and Scopus databases. *Asian Social Science, 9*(5), 18–26.

College Library Services CSN. (n.d.). *Databases vs. Search Engines: What's the Difference?* Retrieved January 8, 2018 from https://www.csn.edu/sites/default/files/legacy/PDFFiles/Library/dbasesearch3.pdf.

Durbin, C. G. (2009). How to read a scientific research paper. *Respiratory Care, 54*, 1366–1371.

Espacenet - Home page. (n.d.). Retrieved March 19, 2018, from https://worldwide.espacenet.com/.

European Patent Office, & European Commission. (2007). *Why researchers should care about patents*. European Commission, (pp. 1–8). Retrieved from http://ec.europa.eu/invest-in-research/pdf/download_en/patents_for_researchers.pdf.

Fink, A. (2005). *Conducting research literature reviews: From the Internet to paper*. Thousand Oaks: Sage Publications.

Google Patents. (n.d.). Retrieved January 8, 2018, from https://patents.google.com/.

Khare, R., Leaman, R., & Lu, Z. (2014). Accessing biomedical literature in the current information landscape. *Methods in Molecular Biology (Clifton, N.J.), 1159*, 11–31.

Kramer, D. (2005). How to read a patent. *ASHRAE Journal, 47*(5), 90–92.

O'Leary, Z. (2005). *Researching real-world problems: A guide to methods of inquiry*. Thousand Oaks: SAGE.

Shah, P. (1998). *Successful study: The essential skills*. Andover: Cengage Learning EMEA.

Tantra, R., Brown, R. J. C., Milton, M. J. T., & Gohil, D. (2008). A practical method to fabricate gold substrates for surface-enhanced Raman spectroscopy. *Applied Spectroscopy, 62*(9), 992–1000.

WIPO - Search International and National Patent Collections. (n.d.). Retrieved January 8, 2018, from https://patentscope.wipo.int/search/en/search.jsf.

Chapter 4
Health and Safety

It is easy to disregard health and safety, especially when you are in the middle of your research investigation and on the verge of a significant experimental discovery. It may be that you are pressed for time or that you simply want to do a quick experiment in the laboratory to see if an idea works. As such, you are tempted to carry out activities without carrying out appropriate assessment of hazards and risks. Yet, by not caring about health and safety, you (as well as others) can be harmed or injured. Hence, in the first part of this chapter, I aim to discuss the importance of health and safety—why you should care and the consequences of not caring. The second part of the chapter will give you practical guidelines, the dos and don'ts that you will need to know before you step into a scientific laboratory. As the topic of health and safety is massive, I will only be covering some of the standard and basic practices that are common to all scientists. Hence, I will cover topics such as general safety, fire safety, handling of general chemicals and their disposal. I will also be covering some of the common forms that you are likely to fill, i.e. risk assessment and Control of Substances Hazardous to Health (COSHH). Please note that this chapter will not cover specific scenarios such as how to deal with lasers, gases, biological agents, ionising radiations, nanomaterials, etc. Also, I will not be dealing with personal circumstances, e.g. those with disability and expectant mums. Overall, my goal for the chapter is to give you a brief overview and so the guidelines presented here should be used as a starting point for further considerations, to be discussed with your colleagues and safety advisor. It may be that you will want to develop them and edit them further, to take into account the circumstances specifically associated with your laboratory. Please note that in some countries such as the UK, health and safety is an integral part of legislation. In such circumstances, it is likely that the organisation in which you work for will already have a system in place and so you must be clear on any ground rules already imposed by the organisation. If in doubt, I suggest that you discuss your situation further with your organisation's designated safety officer.

R. Tantra, *A Survival Guide for Research Scientists*,
https://doi.org/10.1007/978-3-030-05435-9_4

4.1 So, Why Should You Care?

Although accidents (especially near miss) can happen in a laboratory, these are often dismissed and do not get reported. However, there have been certain instances in which the injuries from accidents can be so bad (e.g. leading to deaths) that these eventually attract the attention of media. There are some harrowing examples from the past that I would like to share with you, in order to illustrate my point:

- In 2008, a 23-year-old female research assistant died, due to horrific burns from a laboratory fire. According to the news (Van Noorden, 2011b), she was transferring a pyrophoric reagent, i.e. tert-butyllithium, from one sealed container to another. During the transfer, the plastic syringe fell apart, thus spilling the chemical and igniting a fire. At the time, she was not wearing the correct protective equipment. This particular tragic incident not only led to the death of an innocent life but also prosecution, which was the first ever criminal case that resulted from an academic laboratory accident.
- In 2009, a male graduate student in Texas Tech University in Lubbock had lost three fingers whilst performing a dangerous experiment. He was handling nickel hydrazine perchlorate, which resulted in detonation; at the time he has been reported to have used hundred times the recommended amount (Van Noorden, 2011a).
- In 2011, a female undergraduate working at Yale University's Sterling Chemistry Laboratory was found dead in a laboratory (Van Noorden, 2011a). She was found with her hair tangled in a lathe and died of asphyxiation.

These gruesome cases clearly show that there are serious hazards and risks associated with laboratory activities/practices. It is therefore imperative that you should take the necessary measures to minimise/remove the risks. Remember to assess ALL activities and not to overlook any activities that are considered to be routine or perceived as safe. Let me illustrate this point by sharing with you my own experience, of what happened to me a few decades ago when I was working in microfluidics laboratory. As part of my research I had to dry my glass microfluidics chip after washing it. The drying process usually occurred in an oven, in which the temperature of the oven can be ramped up and down, as appropriate. In my haste one day, I decided to use a hotplate instead and without thinking had put a cold glass chip to a preheated hotplate. The chip exploded! Luckily, the entire process was conducted in a fume hood and so the small explosion had been contained and no one got hurt. Nonetheless the experience had taught me to take more notice of health and safety and not to overlook any activities, including the more simple and straightforward tasks.

4.2 Basic Dos and Don'ts

There are some basic rules that all scientists will need to follow, whether they are working in a chemical laboratory or a biological laboratory. Some of these generic rules are pretty obvious, but sometimes researchers do not even bother to follow such rules. In this section, I present as to what some of these basic rules may be.

4.2.1 General Safety (Bettelheim & Landesberg, 2013; Elias, 2002)

- Wear sensible attire, e.g. full-length trousers rather than dresses or shorts, to avoid exposing any part of your legs. Wear sensible shoes, e.g. no open-toed sandals.
- Do not work alone without anyone within call.
- Do not wear clothing that may get in the way when you are performing an experiment. If you have long hair, make sure you tie it back.
- Familiarise yourself with the meaning of safety symbols, such as toxic/very toxic and corrosive.
- Upon entering a laboratory, put on a laboratory coat, safety goggles and gloves.
- Depending on your activities ensure that you have the right protective equipment for the job. For example, you may need the right type of goggles, a face shield, a full-face respirator or a chemical-resistant apron. If you are moving gas cylinders, you will need to wear steel toe-capped shoes, as well as industrial quality gloves. Ensure that you are wearing the right type of gloves for the chemicals/substances that you are handling; refer to the corresponding MSDS sheets for further information.
- When wearing gloves, do not touch any exposed parts of your body such as your face and mouth.
- Do not eat, drink, smoke or apply cosmetics.
- Do not use fridge and freezer to store food, beverages or personal items.
- Ensure that you have access to antidotes, emergency stations or safety equipments, such as showers, eye wash, first-aid kit and fire alarm. Ensure that these all are functioning properly and that you know how to use them.
- Make sure that you label everything, so that people are aware of the contents in containers; don't leave unlabelled material for disposal by others.
- Do not smell and taste any chemicals or unknown/unlabelled substances. If you come across any unidentified substances, contact your health and safety officer to identify further actions.
- Clear up spillages.
- Clear up after broken glass immediately.
- Before leaving the laboratory, remember to remove your coat, gloves, goggles, etc. Also, remember to wash your hands.

- Report all accidents (including any near misses), no matter how small these might be.

The above guidelines should enable you to start thinking about other dos and don'ts that you may wish to include, as part of general safety, relevant to activities common in your own laboratory. For example, you may want to remind laboratory users to (Spellman, 1998):

- Not pipette anything using their mouth.
- Protect their hands with a thick towel when inserting a glass tube into a rubber or plastic tube.
- Not keep any flammables or organic solvents near a Bunsen burner.
- Take care when using a sand bath, e.g. making sure that there are no residues of any organic compounds or spills still present in the sand.
- Etc.

4.2.2 Fire Safety Practices

Fire prevention is everyone's responsibility and you should be aware of some of the basic safety practices associated. These are important in order to reduce the destruction caused by fire, potentially saving lives. An essential element in fire safety is to understand what causes and what elements sustains a fire (Klinoff, 2013). As such, you may already be familiar with the so-called fire triangle (as illustrated in Fig. 4.1). In this triangle model, you will see three different elements that are needed in order to start and sustain a fire: heat, fuel and oxygen. If any of the element is removed, then you will in effect remove the fire. The removal of heat or fuel from the equation in order to remove a fire is pretty obvious. However, we do not often think much about oxygen as the other factor. It is worthwhile to always remember this third element, as you may be in a position to control or remove a fire if you remember to remove the oxygen from the equation. For example, you can easily contain a fire by remembering to close the door behind you after leaving a room.

If you work for an organisation that offers fire training, then you should try to attend this. The fire training will not only equip you on what you have to do in an event of a fire but will train you on the use of suitable equipment. You will learn that there are different classes of fires, arising from ordinary combustibles, flammables, electrical source or combustible metals. By knowing what kind of fire you are dealing with, you will then be able to identify the correct fire extinguisher to use.

As part of a growing list of dos and don'ts, you may want to include some guidelines that are specially associated with fire safety measures (Perry, 2003):

- Keep things tidy. Dispose of any unnecessary combustible materials in the laboratory, e.g. unwanted cardboard boxes.
- Store chemicals correctly; for example flammables should be stored in the correct designated and approved cabinets.

- Ensure that paths to exit doors are free from obstructions.
- Ensure that you report all fires, no matter how small these may be.

It is important to also have in place a plan on what to do in the event of a fire, in which everyone will need to familiarise themselves with. For example:

- What happens when you discover a fire (Ray, 2007)?
  - If you discover a fire, decide if the fire is small enough to be extinguished. Use equipment such as blankets and fire extinguishers (only if you have received proper training).
  - If you are not able to contain the fire, shout fire to alert those surrounding you. Report the fire immediately to emergency services.
  - Pull the fire alarm upon exiting the building and close the door on your way out.
- What happens if you are on fire (Hill & Finster, 2016)?
  - Recognise that this is likely to be associated with the burning of skin and clothing, which is a class A fire (i.e. combustibles).
  - Understand that you have several options. According to Hill and Finster (Hill & Finster, 2016), you need to stop what you are doing and drop yourself to the floor. Then whilst covering your face, you need to roll around in order to extinguish the flames. Another way to extinguish the flames is to pat the area

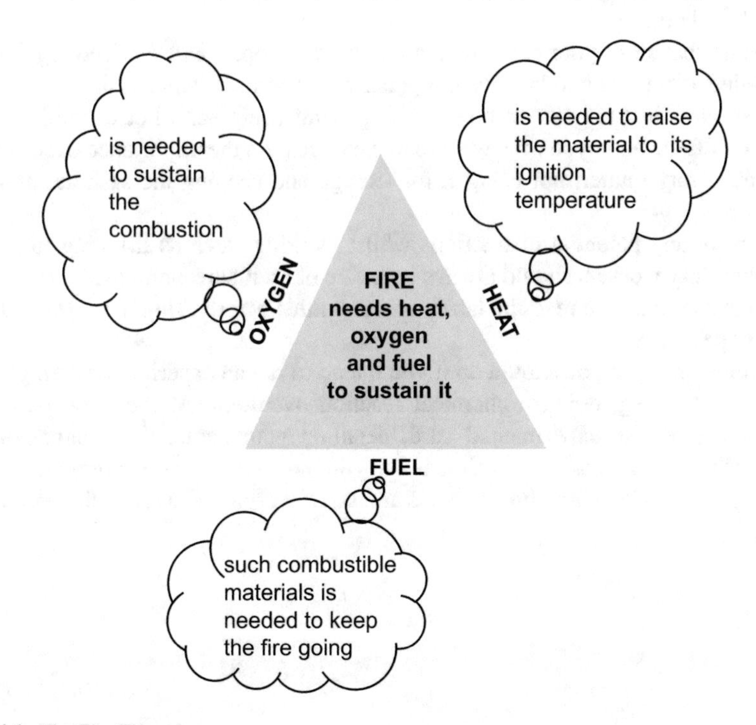

Fig. 4.1 The Fire Triangle

with towel or jacket or fire blanket. Another option is for you to go to safety shower (if there is one close to you). If you choose this last option then you must bear in mind the distance that you will have to run to one. Running to a shower can cause the fire to increase and there is also the possibility that you will be inhaling toxic fumes. So choose the shower option, only if it is accessible, i.e. is within a short distance from you.

4.2.3 Running an Experiment?

If you plan to run an experiment in the laboratory, remember to (Jardine, 1994):

- Carry out only the experiments in which you have been authorised to do so.
- Make sure that you have filled in the relevant safety forms (to be signed/approved by a senior member of staff), e.g. risk assessment and COSHH. Ensure that you are aware of any specific risks associated with your experimental investigations. Take the necessary precautions if you are dealing with ionising radiations, lasers, biohazards, etc. and fill out the appropriate forms. If you are unsure about what forms to fill and how to fill them, then contact your health and safety advisor to further discuss.
- Ensure that all signed forms, such as risk and COSHH, are visible and displayed in the laboratory.
- Ensure that all of your equipment are in working order and serviced regularly if needed, e.g. fume hoods, Gilson pipettes and deionised water system.
- Have clear instructions on how to use general equipment. For example, in the case of fume hood you may want to make it clear on the importance of removing unnecessary clutter, not using it for storage and keeping the sash at approved levels.
- Remove any potential distractions whilst working, such as listening to music. Laboratory workers should always be aware of their surroundings.
- Make sure that you read the labels on the bottles before using them when doing an experiment.
- Be clear on what you should do if you intend to run an experiment when you are not around, e.g. doing a chemical reaction overnight. At the very least, you should place an experimental card, detailing your name, date, nature of the experiment, duration of experiment and your contact telephone number.
- Keep it tidy. Clean up after yourself and store everything away at the end of the day.

4.2.4 Waste Management

Remember, you cannot dispose hazardous wastes in drains, as they can get into the water system, pollute the environment, cause diseases, etc. Ideally, your organisation should have established a waste protocol for you to follow and as such you should be clear on this. Minimally, as part of a waste management strategy, you will need to (National Research Council (U.S.). Committee on Prudent Practices in the Laboratory, 2011):

- Try to minimise the amount of waste that you create, whenever possible. For example, do not over-order on chemicals.
- Decide what you can or cannot dispose down the drain, as governed by legislation. For example, it is sometimes normal practice for a laboratory to dispose non-toxic substances, such as aqueous buffers and common acids/alkalis (with excess water for dilution) into drains.
- Segregate the different types of waste and ensure that you have ways of collecting them in separate containments. Wastes from a laboratory can have different classifications, e.g. general, chemicals and sharps. This can be further complicated by the fact that you can have mixed wastes, e.g. biological mixed with chemical wastes. Decide in advance if it is possible to put several types of wastes in one container. For example, in a general chemistry laboratory, it is likely that you will have waste bottles labelled as unhalogenated and halogenated organic solvents. Remember, it is important to segregate the wastes sufficiently so that they do not react with one another. Some examples of **incompatible** wastes (and thus should **NOT** be mixed) include acids with bases, oxidisers with flammables, water reactive with aqueous chemical and cyanides with acids.
- Ensure that the containers you use are suitable for the type of waste/s. For example, don't store acids in metal containers.
- Store waste bottles in secondary containments, in designated approved areas, e.g. fume cupboard or vented cupboard.
- Label your waste correctly. If chemicals are unlabeled then these become unknown, in which case your organisation may need to send them out for analysis, which can be very costly.
- Process and manage your waste as soon as you finished your experiment. Protocols on how to manage waste should be detailed in your relevant COSHH form.
- Ensure that you do not overfill your waste containers. When a container is full, you will need to dispose of your waste according to the protocol that has been set out by legislation and guidelines provided by your organisation. Your company may organise waste pick-up points, for you to transport your bottles to. If you are transporting bottles of wastes, remember to use a secondary containment and transport them using a stable cart; do not overload the cart.

As a final note in this section, remember that the rules set out above, e.g. in relation to general safety, fire safety, etc., are there for you to consider further. You may need to edit them further in order to suit your particular laboratory requirements. Remember that whatever rules you decide to adopt for your own laboratory, you will need to discuss these with your laboratory colleagues and safety advisor, in order to reach an agreement. Once you have reached a consensus on what the rules should be, you should print them out and paste them on the wall in the laboratory, as a constant reminder to both workers and visitors.

4.3 Assessing Hazards and Potential Risks

Nowadays, the process of assessing hazards and risks is often formalised. As a result, it is common for researchers to fill in forms such as risk assessment and COSHH. By filling out these forms, you will be able to identify the hazards and potential risks arising from any activity that you do in the laboratory. The purpose of such forms is to allow you to come up with action points in order to reduce, limit or remove the risk of such unwanted eventualities, thus allowing you to do safe science. Remember that if you do have an accident, then you will need to report all injuries (as well as any near miss), i.e. no matter how small or insignificant the event may be.

In the UK, filling out such forms has become a legal requirement and thus mandatory. If this is the case, then your organisation should have in place standard templates for you to fill, underlining what information is required. My aim for this section is to give an idea of some of the information that are required when filling such forms.

4.3.1 Risk Assessments

Usually this form will require you to (Erhabor & Adias, 2012):

- Identify the task; task area, i.e. location; who will be doing the task; and how often. Remember to include not only yourself but also other colleagues and visitors to the laboratory.
- Identify hazards (the dangers), such as electrical hazards, chemical hazards and hazards arising from equipment, e.g. using drills.
- Identify potential risk from the hazards. Risk is the probability that the danger will become a reality. For the assessment of risk, you will need to identify who is likely to be harmed, how this is likely to happen and what the likely consequence or severity will be. The risks should be ranked in magnitude from low to high. In order to have an estimate of this, you can adopt the formula magnitude of risk = likelihood + consequence.
- Assess the risk that you have listed. Always list and deal with the highest risk first. This should then be followed by those of medium risk, and then the lowest risks last.

- Identify any control measures that you can implement in order to reduce/limit or remove the risks altogether. Consider using engineering controls (e.g. fume hood or having an interlock system in place), using any personal safety equipment, etc.
- Indicate in the risk assessment if you will be dealing with chemicals or other hazardous substances. If you are handling chemicals, then a separate form, i.e. COSHH form, should be filled (see below for further information).
- Review and update control measures regularly. It is important to appreciate that the risk assessment form is a living document and that you will need to regularly check that the control measures that you have implemented are indeed working. If you find that further actions are required, then you will need to report any new findings and identify further actions, specifically what should take place, by whom and by when.

When filling your risk assessments, make sure that you dedicate sufficient time to fill them properly to identify ALL activities that can pose a potential risk. Here are some of the activities that may be overlooked, simply because the risks associated to such hazards are less obvious:

- Dealing with sharps: Remember that there is a high risk of stabbing yourself with the needle upon its removal from a syringe barrel.
- Using a fume hood (Rayburn, 1990): Remember that there may be a risk that the fume hood is not functioning well, e.g. if it has not been serviced or if there is something that disturbs airflow. It may also be that some laboratory users do not know how to properly use the fume hood, e.g. by not having the sash at the right level.
- Hot glassware on cold surfaces (Hong, Rashid, & Santiago-Vázquez, 2016): When a bottle containing hot liquid is placed on a cold surface, there is a high chance that the glass bottle will explode.

4.3.2 COSHH

Remember to fill out a COSHH form if you are dealing with chemicals or other hazardous substances. Before you plan to order any chemicals, please check with your organisation to find out if the substance that you are proposing to use has not been banned. For example, some organisations that I have worked with in the past have banned the use of hydrofluoric acid. Although COSHH covers most substances, they may not cover certain materials like asbestos.

It is likely that your organisation will already have in place a template form for you to fill in, indicating what information is required. Just to give you an idea, a typical COSHH form usually requires you to identify:

- The chemical or substance.
- The quantity already in storage.

- The process in which you will be handling the chemical/substance: In other words, what are you going to do with it? How much of it will you need? How often are you likely to use it for a particular process, e.g. every day or once a week or once a month?
- Those who are potentially exposed to the risk, e.g. laboratory workers, visitors, cleaners, etc.
- The hazards, i.e. the potential that a substance can cause harm: As such, you will need to refer to the information in the corresponding datasheets or the MSDS (Material Safety Data Sheet) of that chemical/substance, which is usually shipped alongside with the chemical/substance that you have ordered. Take note of any hazard symbols that are displayed on the bottle, e.g. toxic, very toxic, harmful or corrosive. The MSDS should also contain other hazard information, e.g. if the substance is mutagenic, sensitiser, carcinogenic and toxic to reproduction. Please take care when handling substances that are unstable, particularly when there is a risk of a potential explosion such as azides, fulminates and acetylides of certain metals (Martin, 1993). The MSDS is also useful to deduce any physical/chemical characteristic information that can be of interest. For example, for the purpose of waste management, you may want to know if your chemical/substance is water soluble or not. Furthermore, you will want to identify what it is incompatible with, e.g. if it is an oxidising or reducing agent. In addition to hazards associated with any starting materials, you will need to also identify hazards associated to the formation of any new compounds that you may synthesise, as a result of a particular process.
- The potential risks, which is the probability of the hazards becoming a reality: As such, you will need to think of the circumstances in which the hazards can cause you (and others) harm. For example, you will need to state if and how the chemical/substance may enter the human body, e.g. direct contact with hands, inhalation and spillages. Is there a possibility of the substance self-igniting or resulting in an explosion or uncontrolled chemical reaction? If so, under what specific circumstances? For example, remember that when doing an acid dilution, you must add acid to water and **NOT** water to acid.
- The control measures that you will implement, in order to substantially reduce or remove the risks all together: If unsure then you need to take professional advice from an expert in the field or do your own research on what past workers have used in order to minimise/remove potential risks.

When coming up with control measures, you can ask yourself the following questions:

- Can the laboratory/system be modified to reduce the risks? For example you may like to adopt the use of engineering controls, such as fume hoods, exhaust ventilation, placing access restriction and having in place a safety lock mechanism.
- Can I use personal protective equipment? Remember to have the correct personal protective equipment, making sure that you are wearing the right kind of goggles, gloves, face mask, etc. for specific tasks.

- Can I select a safer, alternative chemical or procedure? For example, you may want to buy a bottle of acid that has been diluted rather than having to deal with a more concentrated form.
- How am I going to store the chemical/substance (Pipitone & David, 1991)? You may want to set a maximum storage volume for certain chemicals, such as flammables and highly toxic chemicals. Remember to store like things with like. As such, you may need to have in place designated approved cabinets for different chemicals/substances, such as flammables and corrosives. Within each type of cabinet, you may need to further segregate the different chemicals through the use of secondary containments. For example, in the corrosives cabinet, you will want to store acids separately from bases. In addition, you may want to store mineral acids separately from organic acids. Remember to refer to the MSDS for further information.
- How am I going to dispose of the chemical/substance (Furr, 2001)? Can any hazardous substance (or the by-products) from the experiment be rendered harmless prior to disposal?

From my own experience with chemicals, one of the most difficult tasks when it comes to filling out a COSHH assessment form is when I had to deal with a chemical/substance whose hazard is not so known. For example, in 2009, I had to work on a number of nanomaterial projects and at that time the hazards and potential risks associated with these substances were rather sketchy. My approach, with regard to this lack of information, was to treat all nanomaterials as being extremely hazardous. Where it is possible to do so, I tried to estimate the hazard information of known properties of their congeners and be on the side of caution, e.g. imaging the worst-case scenario (Greenemeier, 2008). So, when I had to deal with carbon nanotubes, I more or less handled them as if they were asbestos.

4.4 Managing Health and Safety in a Laboratory

An important aspect of health and safety is to have an effective management system in place and to ensure that all staff and visitors to the laboratory are clear on how things are managed. It is best to summarise any management process via a workflow diagram, which is a graphic representation of the specific process. Figure 4.2 is a workflow overview example on how a laboratory can be managed. The workflow highlights key points and shows you who is responsible for what. As indicated in the diagram, it is clear that it is the job of the laboratory manager who ensures that the proposed health and safety infrastructure is working properly.

You can also have workflow diagrams for other processes as well, e.g. waste management, purchase order management and what to do in case of spillages. Everyone who enters the laboratory should be made aware of such forms. As such you can paste these on the wall of your laboratory, where they can be visible to all.

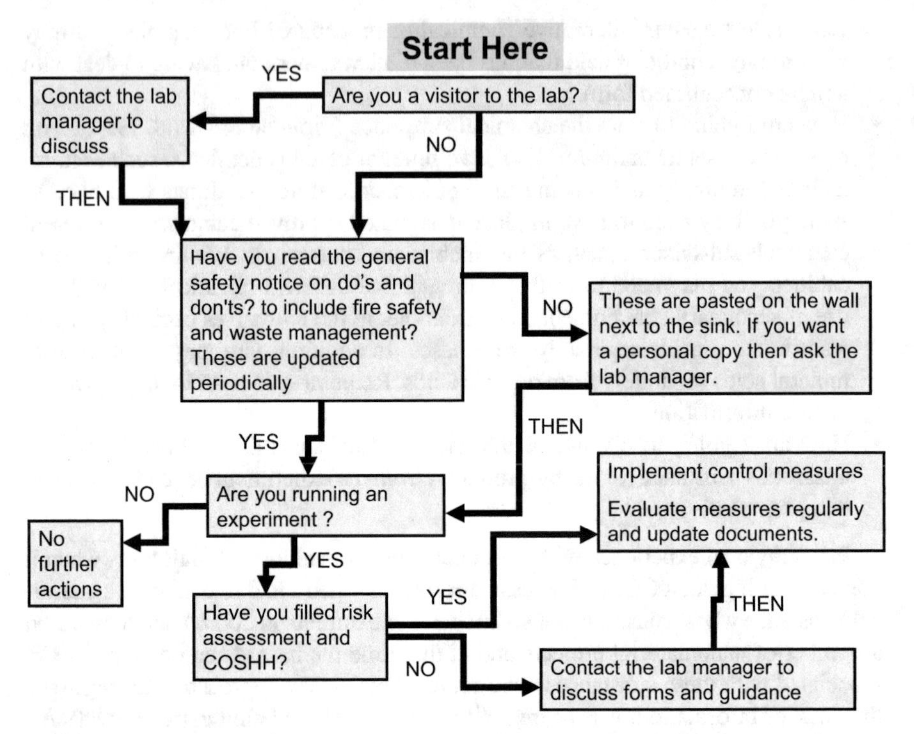

Fig. 4.2 Illustration of a workflow example, clearly showing who is responsible for what in relation to health and safety

4.5 Summary

The purpose of having a health and safety initiative in place is to ensure that your activities in the laboratory will not pose any injury or harm to you, others and the environment. Although the topic of health and safety is huge, my aim here is to give you introduce you to the basics. A key aspect of health and safety is to be aware of the hazards and risks and subsequently implement actions to minimise or remove the risks. It is important to review your safety measures on a regular basis, update them if necessary to ensure that control measures that already in place are indeed effective. Finally, throughout the chapter, I have listed some generic dos and don'ts that everyone who enters the laboratory should adhere to. I have given you some tips with regards to generic practices (e.g. general chemical handling, how to manage wastes, fire safety) throughout the chapter for further considerations.

References

Bettelheim, F. A., & Landesberg, J. M. (2013). *Laboratory experiments for introduction to general, organic and biochemistry*. Boston: Brooks/Cole.

Elias, A. J. (2002). *A collection of interesting general chemistry experiments*. Hyderabad: Universities Press.

Erhabor, O., & Adias, T. C. (2012). *Laboratory total quality management for practitioners and students of medical laboratory science*. Bloomington: AuthorHouse.

Furr, A. K. (2001). *CRC handbook of laboratory safety*. Boca Raton: CRC Press.

Greenemeier, L. (2008). *Study says carbon nanotubes as dangerous as asbestos*. Scientific American. Retrieved September 25, 2017, from https://www.scientificamerican.com/article/carbon-nanotube-danger/.

Hill, R. H., & Finster, D. C. (2016). *Laboratory safety for chemistry students*. Hoboken: John Wiley & Sons.

Hong, S.-B., Rashid, M. B., & Santiago-Vázquez, L. Z. (2016). *Methods in biotechnology*. Hoboken: Wiley-Blackwell.

Jardine, F. H. (1994). *How to do your student project in chemistry*. UK: Chapman & Hall.

Klinoff, R. (2013). *Introduction to fire protection and emergency services*. Burlington: Jones & Bartlett Learning, LLC.

Martin, W. F. (1993). *Protecting personnel at hazardous waste sites*. Amsterdam: Elsevier Science.

National Research Council (U.S.). Committee on Prudent Practices in the Laboratory. (2011). *Prudent practices in the laboratory: Handling and management of chemical hazards*. Washington D.C.: National Academies Press.

Perry, P. (2003). *Fire safety questions and answers: A practical approach*. London: Thomas Telford.

Pipitone, D. A. (David A. (1991). Safe storage of laboratory chemicals. Wiley Hoboken.

Ray, B. (2007). *Modern methods of teaching chemistry*. New Delhi: A P H Publishing Corporation.

Rayburn, S. R. (1990). *The foundations of laboratory safety: A guide for the biomedical laboratory*. New York: Springer.

Spellman, F. R. (1998). *Safe work practices for the environmental laboratory*. Lancaster: Technomic Publishing Company Inc.

Van Noorden, R. (2011a). A death in the lab. *Nature, 472*(7343), 270–271.

Van Noorden, R. (2011b). Chemist faces criminal charges after researcher's death. *Nature*.

Chapter 5
Experimental Design

So, why do you want to conduct an experiment in the first place?

Quite simply, your job as a research scientist is to find answers to questions and in order to do this you will need to design and run experiments.

When I started my PhD. in 1995, I must admit I did not have a clue on how to design my experiments. Like most researchers at the time, I was never formally taught this. So, how did I manage to finish my PhD. within 3.5 years? Quite simply, I was lucky to have been surrounded by experienced postdocs who provided me with the much-needed guidance. As the years went on, I learned that the process of experimental design can be formalised, in which there are several strategies to choose from.

My intention for this chapter is to give guidance to scientists who have not been properly taught (or those in need of a refresher course) on the topic of experimental design. As such, the chapter starts off by discussing the different phases associated with running an experiment. The second part of the chapter is dedicated to understanding two common experimental strategies that you can adopt when running an experiment:

1. How to run a control experiment.
2. How to adopt a design of experiment (DOE) approach.

5.1 The Five Phases of Running an Experiment

Figure 5.1 depicts the five different phases involved should you decide to run an experiment:

R. Tantra, *A Survival Guide for Research Scientists*,
https://doi.org/10.1007/978-3-030-05435-9_5

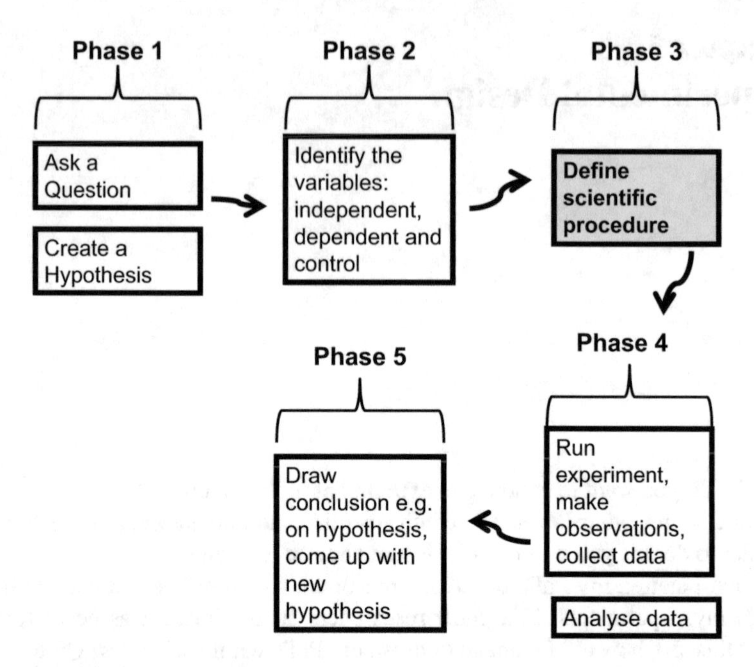

Fig. 5.1 The five phases of running an experiment

1. Phase 1: The aim of this phase is to ask a question and create a hypothesis (Barnard, Gilbert, & McGregor, 2007). Remember that your question must be clear and testable. By testable, I mean that you can perform the necessary measurements during the study and that there is a way in which you are able to control certain conditions of your experiment. After forming the question, you will then need to state your hypothesis. A hypothesis can be thought of as a predication of what is likely to happen as a result of the proposed study. It often constitutes having an IF/THEN statement. In other words, IF a particular scientific phenomenon is true THEN certain observations are expected. In the formation of a hypothesis, you must have a credible scientific explanation as to why the scientific phenomenon that you are predicting may hold true. It is likely that you can generate several hypotheses from a given question. However, it is important to remember that when carrying out an experiment, you can only test one hypothesis at a time.
2. Phase 2: The aim of this phase is to identify key variables. In simple terms, variables are those factors in an experiment that can affect its outcome. There are different types of variables that you should be familiar with, namely (Williamson & Bow, 2002):

   - Independent variables: These are the factors that you are changing in an experiment, chosen specifically in order to support or disprove your hypothesis.
   - Dependent variables: These are the factors that you are trying to measure, as a result of your independent variable. Although dependent variables should

predominantly be quantitative in nature, it is possible to have qualitative dependent variables (which may help with data interpretation).
- Control variables: These are the factors that you should keep constant throughout the experiment.
- Noise (or random) variables: These are variables in which you are not able to control or manipulate. Examples include fluctuations in ambient temperature, humidity, quality of raw materials, etc. (Khuri & Cornell, 1996).

3. Phase 3: The aim of this phase is to define your scientific procedure, i.e. a step-by step instruction on how you are planning to conduct the experiment. In defining your experimental procedure, it is important to take into account the original question, i.e. hypothesis and all of the key variables previously identified. In order to prove or disprove your hypothesis, your step-by-step instruction will need to include information on:

   - The level of treatment of the independent variable, in which you will need to define a range. Within this range you will need to include different scenarios of interest. You will need to include the CONTROL GROUP in your experimental design, i.e. data when acquired under normal conditions (Ary, Jacobs, Razavieh, & Ary, 2010). Please remember that the CONTROL GROUP IS NOT THE SAME AS YOUR CONTROL VARIABLES (as discussed in Point 2). You will also need to identify conditions of interest that are on either side of the control group. For example, you may want to simulate real-world conditions or you may be interested to see what happens at the lower or higher end of the range.
   - Level of replication associated with the dataset generated (Ellison, Farrant, & Barwick, 2009): It does not make scientific sense doing an experiment once and never being able to repeat it. Therefore, you will need to decide how many times you will need to repeat a certain procedure. This is necessary for the purpose of data reliability, thus verifying that any interesting observations are not solely due to the normal variations expected in a given data set. As a general rule, you should choose to run six replicates.

   Remember that when coming up with the scientific procedure, you should always try to adopt best practice methods (such as using a reference material to calibrate your instrument prior to its use), whenever possible. This is important to minimise potential measurement errors that can affect your final results please refer to Chap. 6 on the topic of best practice, for further information.

   Remember to document any experimental procedures that you are using and to meticulously record the details straight into your laboratory notebook. If at any time you have made any changes to the initial scientific procedure, remember to note them down and ensure that an explanation is given for such changes.

4. Phase 4: In this phase, you will actually be carrying out the experiments, making observations. You will need to collect your data for subsequent analysis. As you may be churning out a lot of data, you will need to decide what information you want to extract, in order to subsequently carry out the necessary statistical analysis to prove or disprove your hypothesis. Data treatment can vary, from adopting a

very simple statistical approach such as calculating standard values (e.g. mean and standard deviation) to the use of elaborate statistical techniques such as principal components analysis (PCA). PCA is particularly useful for handling large data sets and will allow you to visualise the data better (thus enabling you to compare different data sets with ease) (Brereton, 1992; Jolliffe, 1986). I will not go into details as to what statistical analysis you will need to carry out, as this will be dependent on the kind of study that you are conducting in the first place. For more information on how to treat your experimental data, please read Chap. 6 (Best Practice). For further information on to statistics and data analysis please refer to the relevant literature, e.g. Cowan, (1998), Diggle and Chetwynd (2011) and Livingstone (2009).
5. Phase 5: The aim of this phase is to draw conclusions from your data analysis. At this point you can either accept or reject your hypothesis. You may also come up with new hypothesis to test out at a later date.

Out of all the five phases, it is phase 3 that will be key to your success as this phase will determine data quality. As such, the remainder of this chapter aims to discuss phase 3 further. There are two common strategies that you can adopt for phase 3, namely:

1. Strategy 1: running a control experiment.
2. Strategy 2: adopting a design of experiment (DOE) approach.

5.2 Strategy 1: How to Run a Control Experiment

Scientists often adopt this basic strategy when they are starting out in research. It is a simple approach as the experimental design involves varying one independent variable at a time (Kirk, 2012) whilst keeping all other variables constant (apart from the random variables, in which you have no control of). If you happen to unintentionally vary more than one (independent) variable at a time, then this will invalidate your findings when using this kind of approach.

In order to illustrate the steps involved in a control experiment, I would now like to present a case study. This case study surrounds an experiment that aims to improve the repeatability of a technique called surface-enhanced Raman scattering (SERS). But before delving into this, I will be giving you some background information on this analytical technique.

Background to the case study: SERS

SERS is an extended form of the well-known and widely accepted Raman spectroscopy technique. Like infrared spectroscopy, the power of Raman lies in its ability to give a spectral fingerprint (unique for that sample). However, a major issue with Raman is the need to have highly concentrated samples, in order to collect the inherently weak Raman signal. The SERS technique has the advantage of allowing the Raman spectral fingerprint to be amplified, thus enabling a much lower detection level to be achieved (typically in the sub-micromolar concentrations). In fact, with some analytes (e.g. Rhodamine 6G), the SERS technique can result in a signal enhancement of 10^{14} to 10^{15} level, thus pushing the detection limit into the realms of single-molecule detection (Royal Society of Chemistry (Great Britain). Faraday Division, 2006). Despite its promises however, SERS has never reached commercial success and its uptake into industry has been poor. The problem has always been associated with poor repeatability.

In 2006, I was investigating ways to optimise the SERS technique, in order to improve its repeatability. At the time, the substrates used for SERS often involved the formation of metal clusters; these were formed using an aggregating agent and synthesised metal colloids (as depicted in Fig. 5.2). Quite common was the use of silver or gold metal colloids with potassium chloride as the aggregating agent. The analyte of choice for most researchers in SERS is Rhodamine 6G, as this analyte resulted in a very high signal enhancement factor during analysis.

So, if you were asked to run a control experiment to investigate ways to improve the repeatability of SERS, how would you do this? See below for details on what you can potentially do.

Designing a Control Experiment on SERS

1. *Ask a question.*

   Can the repeatability of the SERS analysis be improved by changing vortexing time during the metal cluster formation?

   (Note: Remember that the question must be clear and testable, i.e. measurable and controllable.)
2. *Form a hypothesis.*

   IF I optimise the vortexing time during metal cluster formation stage, THEN the metal clusters formed will potentially be of similar size and shape. This has the effect of improving SERS repeatability.

   (Note: Remember that the hypothesis relates to an IF/THEN statement, which must be linked to the question in the first place. Also, you must have a credible scientific explanation as to why the scientific phenomenon that

(continued)

you are predicting may hold true. In this case, I have used findings from my literature search to make the prediction.)

3. *Identify the variables, i.e. independent, control and dependent.*

   Independent variable: Vortexing time during metal cluster formation (units in seconds).

   Dependent variable: SERS signal (units: wavelength, intensity), background fluorescence signal and SERS signal.

   Control variables: operator, type and source of metal colloids, temperature of starting material, stock solution concentrations, type of Raman spectrometer instrument used, type of container used to handle sample, type of vortexing equipment used, Raman instrument setting, etc.

   (Note: Remember that when it comes to designing a control experiment, you can only study one independent variable at a time! In this case, I have chosen the independent variable to be the time needed to vortex the mixture prior to analysis, i.e. vortexing time. I have chosen this as I believe that this factor is the most influential.)

4. *Design the procedure.*

The steps are as follows.

(*Note that the experimental method below is based on previously published work* (Tantra, Brown, & Milton, 2007).)

(a) Operator: Ratna Tantra.
(b) Synthesise silver colloids via Lee and Meisel method.
(c) Prepare stock solutions of Rhodamine 6G (0.01 mM, 10 mL) … etc.
(d) Calibrate Raman instrument using Si, … etc.
(e) Container used: 2 mL Plastic vials. Quartz cell used: Demountable quartz cell (with path length of 1 mm) Hellma, UK. Cleaning step for quartz cell: 2% Hellmanex, deionised water for rinsing … etc.
(f) Vortexing equipment: Fisherbrand Mixer/Vortex Whirlmixer Plus … etc.
(g) Mixing step using the stock solutions: Aliquot 40 μL of potassium chloride to 440 μL of deionised water. Vortex for 2 s. Then add 20 μL of Rhodamine 6G to this … etc. **Vortex for X seconds (in which $X = 0, 10, 30, 60$ and 90 s).** Aliquot 80 μL of the mixture. Place in pre-cleaned quartz cell. Acquire data after a 2-min incubation time.
   (Note: The statement in bold text is the independent variable of the study. The 30-s case is considered to be the norm and thus identified as a control group.)
(h) Data acquisition: Raman instrument setting: laser excitation (0.5 mW), objective lens x20 … etc. Six times replicate measurements taken per sample and then take average spectrum … etc.
(i) Data processing.
   Part 1: SERS signal (wavelength, intensity), extract mean and standard deviation of main Rhodamine peaks.

(continued)

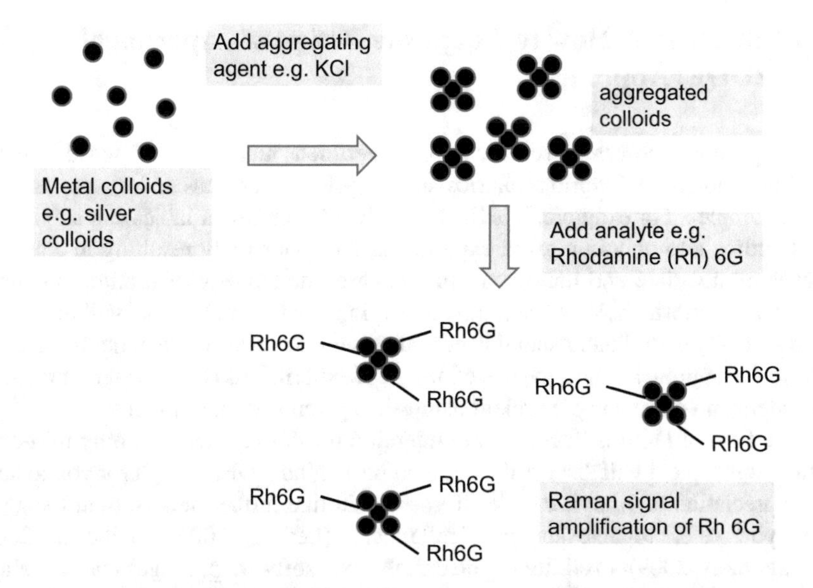

Fig. 5.2 The concept of conducting a SERS analysis. In this example, Rhodamine 6G (Rh 6G) adsorbs on to the metal clusters in order to achieve the SERS signal amplification

> Part 2: Fluorescence intensity associated with Rhodamine peak at ~1700 cm^{-1}; extract mean and standard deviation.
>
> Part 3: Subtract fluorescence background from all SERS data, and then input all spectra into principal component analysis (use Simca-P version 11, Umetrics AB software package); plot principal components against one another to find patterns of data clusters … etc.

Overall, when conducting a control experiment, remember to:

1. Identify your independent, dependent and control variables.
2. Identify the control group, which is DIFFERENT FROM THE CONTROL VARIABLES.
3. Add the level of treatment on either side of the norm.
4. Define the extent of the replication treatment.
5. Define how you are going to process the data.

Ideally, the procedure written in the above case study would have contained more details. For example, with point 14, I can elaborate how the fluorescence background from the data should be carried out.

5.3 Strategy 2: How to Adopt the Design of Experiments (DOE) Approach

Although the approach to run a control experiment makes perfect sense, often it does not capture real-world scenarios, as the system or process under study may be more complex. For example, it is likely that two (or more) variables can simultaneously come into play in a given experiment, thus potentially resulting in a system that is multivariate and non-linear. In this case, the strategy of running a control experiment, in which you are effectively varying one factor at a time, will inevitably give you only partial understanding of your system. Instead, a researcher must adopt a different approach. Thus, the use of design of experiments (DOE) is recommended to understand more complex relationships in a given system or process.

The field of DOE is huge and my intention for this chapter is to only introduce you to this topic. I will discuss the main points behind DOE, enough for you to have an appreciation of what it can do. If you would like further details, then I suggest that you refer to relevant past references (Levine, 2006; Mathews, 2005; Montgomery, 2009). In addition, you can also buy software packages that will allow you to conduct DOE with greater ease, making DOE more accessible to scientists. It is always a good idea to talk to well-known DOE software developers such as Minitab and JMP and Statistical Discovery from SAS ("Minitab Statistical Software - Minitab", n.d.) ("JMP 30-Day Trial Download | JMP", n.d.).

So how can you use DOE to design and analyse your experiments?

In general, there are six key steps that are involved in the DOE approach:

1. Acquire as much knowledge as possible about the system or process under study beforehand. This knowledge may come from data that you have gathered as a result of earlier experiments, historic information from other workers, understanding the fundamental principle or theory of your system, etc. I am stressing here on the importance of being thorough with your background research; otherwise doing a DOE can potentially be a wasted effort.
2. Identify two things: The response that you are measuring (e.g. your dependent variables) and other important factors (e.g. independent variables).
3. Be clear on your goal and what you want to achieve. What do you want to learn from running a DOE?
4. Use your knowledge (from point 1) to design an experiment and identify the factors (from point 2) that are governing your experiments. Choose the best ranges over which to vary certain governing factors.
5. Write the DOE scientific procedure and subsequently run your experiment to collect data.
6. Statistically fit resulting data into an equation (usually a polynomial in the factors). This is necessary in order to build an empirical model for the scientific system or process, which will subsequently allow you to unearth new information about your process or system, so you can plan for your next DOE. In other

words, the outcome of the first DOE study should help you identify what factor/variable settings to focus on next.

An important aspect of DOE is to follow a cycle of designing, measuring and analysing your data in order to unearth new information, i.e. to repeat steps 1–6 over and over again. The cycle should be repeated until you have reached your goal (point 3), in unearthing the kind of information that you are hoping to get by running a DOE.

In order to illustrate the concept of DOE, let us revisit the same SERS case study from before.

So, if you were asked to run a DOE to investigate ways to improve the repeatability of SERS, how would you do this?

Designing a DOE on SERS

1. *Define the goal*

   To investigate the optimum conditions that will result in repeatable SERS measurement.
2. *Define the response*

   I will need to rely on my knowledge. For example, from the literature, I know that the repeatability of the SERS signal is governed by the formation of repeatable metal clusters (in relation to their final size and shape). As it is difficult to reliably monitor the shape of particles in a polydisperse colloidal metal suspension, the focus here has to be on particle size measurements. In particular, it is the polydispersity index (PDI) values (acquired from a particle analyser instrument) that I will be interested in. Hence, the PDI value from such an experiment will be my chosen response.
3. *Define what you want to see in the response*

   Ideally, I will want the PDI response to be at a minimum; the response should be measured after the addition of an aggregating agent to Rhodamine 6G and metal colloids. The smaller the PDI measured, the more homogeneous the metal clusters will be, which technically should improve SERS repeatability.
4. *State what you want from the DOE*

   I will want to know the factor (i.e. variable) settings that will result in small PDI values when doing a SERS analysis.
5. *Identify the factors/variables that can affect the SERS response*

   Again, I will need to rely on my knowledge. According to literature, I have identified a number of factors that can affect repeatability of SERS, to include vortexing time, type of metal colloid used, type of aggregating agent used, concentration of aggregating agent, etc. I will need to assess this exhaustive list and shortlist, i.e. to anticipate the main governing factors that will significantly affect the analysis.

(continued)

6. *Anticipate possible interactions between different factors/variables*

   When carrying out this step, I will need to realise that several factors may interact with one another. For example, from my knowledge of chemistry and SERS, I suspect that strong interactions will occur between the following factors: vortexing time and concentration of aggregating agent. As such my experimental design can involve changing vortexing time and concentration of the aggregating agent at the same time.

7. *Design the experiment (using a suitable software)*

   Using a suitable DOE software package, I will need to plug in the chosen factors and their associated ranges. As in the case study presented in Sect. 5.2, I will also need to write out the step-by-step procedure before running the experiment.

8. *Conduct the experiment in the laboratory*

   Remember that I will need to do this in accordance to the chosen scientific DOE procedure, as detailed in point 7.

9. *Analyse the data collected*

   Using a suitable DOE software, I will need to build and check empirical models from the results, e.g. by fitting a second-order polynomial to the data points.

10. *Repeat the cycle, starting from point 1*

    I will need to use knowledge gained from my last experiment to design another DOE experiment and repeat process with a new procedure. In other words, each time I gain additional knowledge, I will need to repeat the DOE cycle, until I have achieved my goal. In my SERS case, I will want to come to a point in which I am able to identify the optimal conditions for me to produce a small PDI response. For example, from early experiments I may have identified those factors that do not hugely affect the response, e.g. type of storage container used. As such, in my next DOE design, I will want to purposely omit such factors. This will allow me to focus on the interactions between key factors that are of greater significance.

11. *Confirm/validate findings*

    Once I know what experimental conditions will result in minimum PDI value, I will need to verify my findings; that is, I will need to conduct SERS repeatability experiments to confirm.

5.4 Summary

In this chapter, I have discussed the process (consisting of five phases) of what happens when you run an experiment. Out of the five phases, it is the third phase, i.e. when you need to define the scientific procedure, that will be key to your success in the laboratory. In terms of experimental design, there are a number of strategies that you can adopt. I have discussed two of the most common strategies involved in experimental design, namely:

1. How to run a control experiment.
2. How to run a DOE.

Out of the two, designing a control experiment (in which you study one independent variable at a time) is the simplest. However, such a simplistic approach can be a naive way of doing your experiments, as you may not be able to capture real-life scenarios. In reality, scientific processes and systems are usually more complex, in which the measured response can be of multivariate in nature, e.g. when you have two (or more) independent variables interacting with one another. If there are cross-terms, then it is likely that your data will not be linear. In such a case, a different approach to experimental design is needed, in which the use of the DOE approach is recommended. The power of DOE lies in the combination of integrating your knowledge (about the system/process under study) with applied statistics. In order to illustrate the thought process behind the two different approaches to experimental design, I have used the SERS case study to exemplify.

References

Ary, D., Jacobs, L. C., Razavieh, A., & Ary, D. (2010). *Introduction to research in education*. Boston: Cengage Learning.

Barnard, C. J., Gilbert, F. S., & McGregor, P. K. (2007). *Asking questions in biology: A guide to hypothesis-testing, analysis and presentation in practical work and research*. London: Pearson Education.

Brereton, R. G. (1992). *Multivariate pattern recognition in chemometrics: Illustrated by case studies*. Amsterdam: Elsevier.

Cowan, G. (1998). *Statistical data analysis*. Wotton-under-Edge: Clarendon Press.

Diggle, P., & Chetwynd, A. (2011). *Statistics and scientific method: An introduction for students and researchers*. Oxford: Oxford University Press.

Ellison, S. L. R., Farrant, T. J., & Barwick, V. (2009). *Practical statistics for the analytical scientist: A bench guide*. Cambridge: RSC Publishing.

JMP 30-Day Trial Download | JMP. (n.d.). Retrieved October 2, 2017, from https://www.jmp.com.

Jolliffe, I. T. (1986). *Principal component analysis*. New York: Springer.

Khuri, A. I., & Cornell, J. A. (1996). *Response surfaces: Designs and analyses*. New York: Marcel Dekker.

Kirk, R. E. (2012). *Experimental design: Procedures for the behavioral sciences*. Thousand Oaks: Sage Publications.

Levine, D. M. (2006). *Statistics for six sigma green belts: With Minitab and JMP*. London: Pearson P T R.

Livingstone, D. (2009). *A practical guide to scientific data analysis*. Hoboken: Wiley.

Mathews, P. G. (2005). *Design of experiments with MINITAB*. Milwaukee: American Society Quality, Quality Press.

Minitab Statistical Software - Minitab. (n.d.). Retrieved October 2, 2017, from http://www.minitab.com.

Montgomery, D. C. (2009). *Design and analysis of experiments*. Hoboken: Wiley.

Royal Society of Chemistry (Great Britain). Faraday Division. (2006). *Surface enhanced Raman spectroscopy: Imperial college, London, 19–21 September 2005*. Piccadilly: RSC Publications.

Tantra, R., Brown, R. J. C., & Milton, M. J. T. (2007). Strategy to improve the reproducibility of colloidal SERS. *Journal of Raman Spectroscopy, 38*(11), 1469–1479.

Williamson, K., & Bow, A. (2002). *Research methods for students, academics and professionals: Information management and systems*.

Chapter 6
Best Practice

Now, what do I mean by the concept of best practice?

Best practice is often established through trial and error and thus encapsulates the very heart on how best to do things. It can be thought of as the generally accepted guidelines on processes and activities.

As such, whenever you decide to run an experiment in the laboratory, you will need to ensure that you adopt the approach of best practice (whenever possible). This is important, so that you can work more efficiently, thus improving your chance of success, e.g. achieving a more positive outcome from running an experimental study (Osborne, 2008).

This chapter is a natural continuation from Chap. 5 (Experimental Design). As in the last chapter, the aim is to provide you with basic information. When it comes to best practice you must appreciate that different scientific disciplines will have their own set of best practices. As such, it is not the intent of this chapter to deal with best practice that covers all disciplines and eventualities. However, there are some best practices that all experimental scientists should be familiar with, whatever their discipline. As such, the chapter will specifically focus on how you can:

- Develop methods.
- Validate methods.
- Be certain of measurements.
- Record observations into a laboratory book.
- Use statistics to analyse data.
- Ensure quality.

You can use the information presented in this chapter as a starting point to develop your own best practice protocols. Please edit if necessary, in order to suit your personal situation/circumstance.

R. Tantra, *A Survival Guide for Research Scientists*,
https://doi.org/10.1007/978-3-030-05435-9_6

6.1 Method Development

As an experimental scientist, it is likely that you will be asked to develop a method at some point in your career.

So, how can you do this?

For the sake of simplicity here, let's take an example. Let's says that you have been asked to develop a method for the analysis of a particular analyte in a sample matrix. In order for you to do this, there are 5 key steps that you should be aware of:

1. Clarify what you want to achieve from the analysis. As part of this process, you will need to clearly identify your analytical requirements. You will need to identify/clarify your (Zhou, 2011):

   - Sample type: Is your sample biological or chemical in nature, etc.
   - Sample matrix: Are there any potential interferences that can arise from the matrix itself.
   - Sample size: How much sample do you have for the analysis, etc.
   - Data requirements: Do you want a qualitative or quantitative measurement output.
   - Instrument performance requirements: What kind of instrument are you looking for, with regard to sensitivity, selectivity, analysis time, ease of use, etc.
   - Method robustness: Do you want a robust method, i.e. one that will not be sensitive to different operators or day-to-day variability (Ahuja & Dong, 2005).
   - Level of precision and accuracy: What level of accuracy or precision are you looking for in a method? Note that at this point I would like to point out the difference between the two terms, i.e. between accuracy and precision. Let us say that you are taking repeated measurements during an experimental run. Precision is a measure of spread in the given data set. Accuracy however refers to how close the data is to the true value. If you want to achieve the highest level of accuracy in your measurements, then you will need to quantify uncertainties associated with your measurements and possibly achieve measurement traceability (please refer to section below, for further details).

2. Do a thorough literature review (see Chap. 3, Literature Review) and talk to experts in order to rely on relevant (existing) information. An important outcome from your initial research is to identify any potential methods that can be used — to be adopted as is or further develop to suit your particular analytical requirements. For example, if an analytical technique is insufficiently sensitive, then you can edit the method in order to incorporate a pre-concentration step during the sample preparation stage, if possible.

   The source of your documented material can vary, e.g. peer-review articles, books and grey literature. By grey literature, I mean that the documents have not gone through any formal review process to assess for quality. An example of this may be a report from a scientific project, in which the outcome has not been published.

You should also identify any relevant standard documents, to see if there are any useful information that you can use, e.g. on sample preparation. As a side note here, it is worthwhile to highlight that the information from standard documents is usually of high quality. Unlike other types of documents, the development and publication of standard documents will have gone through an extremely formal process, often involving technical committees, working groups, etc. (Murphy & Yates, 2009). In fact, some standard documents such as International Organization for Standardization (ISO) and European Committee for Standardisation (CEN) can only be formed through the involvement of official standard organisations, usually proceeding through government-recognised National Standard Bodies (NSBs), e.g. British Standard Institute (BSI) in the UK. As a result, the publication of a standard document can take years, typically 1–3 years (Hunt, Robitaille, & Williams, 2008).

3. Develop and assess your draft method. Before going into the laboratory and testing out your method, you will need to review every single step in your draft method, to identify where errors can come into the measurement. It may not always be easy to identify sources of errors, as often it is the simplest of things that can contribute to the largest error. Please note that errors are very different from silly mistakes, i.e. human errors. Human errors can include things like taking the wrong bottle and analysing the wrong sample. Undoubtedly mistakes in the laboratory should be avoided but these mistakes are more straightforward to handle, unlike measurement errors.

   So, what do I mean by measurement errors?

   There are two types of measurement errors that you should be aware of (Kozak, 2008):

   - Systematic errors (Myers, 2003): These are associated with errors that can be repeated in an experiment. Some examples include not calibrating your instrument properly, contaminating your sample, destroying your sample during analysis, etc.
   - Random errors (Myers, 2003): These types of errors cannot be repeated in an experiment, being both unpredictable and unavoidable. Typical random errors can be associated to the nature of your instrument and your surroundings, such as variations in ambient temperature and pressure. For example, a common random error in a Raman measurement is often linked to the appearance of cosmic rays in a spectrum, which are automatically picked up by the highly sensitive detector employed in the instrument (Lewis & Edwards, 2001).

4. Upon identifying any potential errors that may arise from your method, you will need to minimise these or remove them altogether. Again, it is worthwhile to conduct a literature review and rely on historic information whenever possible, to find ways to reduce/remove errors from your draft method. However, there are some common practices that researchers should adopt in order to reduce measurement errors. One way is to ensure that your method incorporates the use of reference materials (RMs).

RMs are substances that are employed as a measurement artefacts. RMs are particularly useful to (Zschunke, 2013):

- Detect any fault in an instrument, e.g. if you have incorrect control settings.
- Allow adjustments to be made to the instrument, i.e. calibrate the instruments.
- Train researchers to use a particular instrument; this is an essential part of internal quality checks on operator performance.
- Help to carry out method validation, e.g. used in round robin studies (to compare laboratory performance) or in the development of uncertainty budgets.

As a side note here, it is important to highlight that RMs are often made using high-quality material. For a material to be classified as a RM, it needs to satisfy certain requirements, such as the need to be:

- Sufficiently homogeneous
- Highly stable in relation to one or more specified properties (Zschunke, 2013)

The development of RMs is not trivial and often very costly (National Research Council (U.S.). Committee on Reference Materials for Ocean Science, 2002). As an approximation, RMs require several years to develop before they can be sold commercially. During their development, RM developers will need to:

- Run extensive stability studies.
- Develop suitable homogenisation techniques.
- Run extensive homogeneity trials.
- Investigate packaging and storage condition requirements.
- Identify cost of preparation.
- Take into account marketing considerations.
- Etc.

It is important to bear in mind that not all RMs are of the same quality. There is in fact a hierarchy associated to the quality, which is dependent on how the RMs have been developed and produced in the first place. Certified reference materials or CRMs are of the highest level. CRMs are different from RMs in that CRMs are produced under stricter guidelines. For an RM to become a CRM there needs to be additional evidence related to improved measurement accuracy, often requiring some form of traceability to be established in the measurements (Dulski, 1996). As a result, CRMs will come with a certificate to confirm that they have been produced under the necessary stringent guidelines.

So, how do you know what RMs (or CRMs) are available for use?

There are various sources that you can access to find out, namely:

- COMAR: This is an international database for certified reference materials ("Bundesanstalt für Materialforschung und -prüfung (BAM) - RRR - COMAR Overview", n.d.).

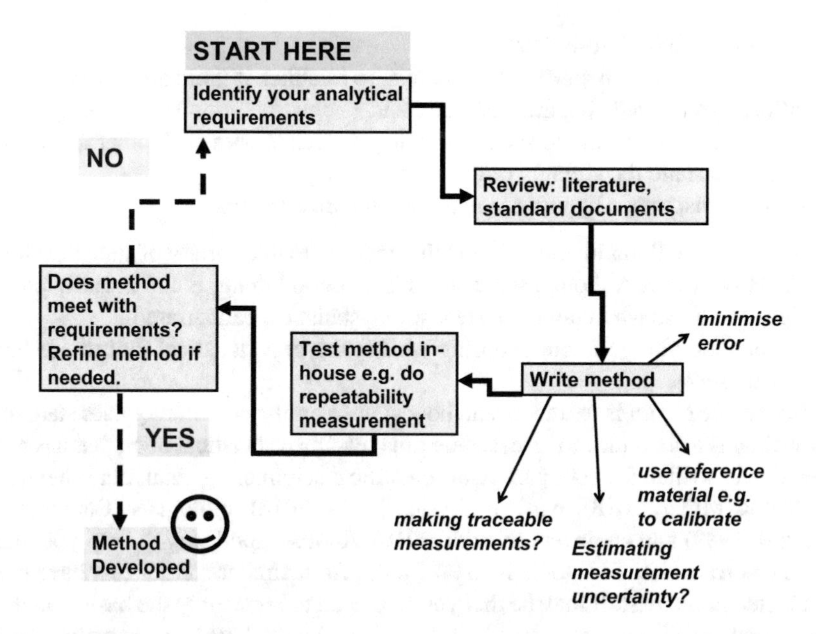

Fig. 6.1 Method development cycle

- Reference material producer catalogues and Internet websites such as NIST and LGC (Walker et al., 2007).
- Specialised technical books: For example, my book "Nanomaterial Characterization: An Introduction" gives a list of RMs and CRMs specifically associated with nanomaterial research (Tantra, 2016).

5. Remember that the process of method development may not be straightforward. It may be that you need to test out the method in the laboratory several times whilst editing the method as you go along (as depicted in Fig. 6.1). This should be repeated until the method developed satisfies the analytical requirements (initially identified, as discussed in point 1).

6.2 Role of Statistics in Data Analysis

The field of statistics is huge and it is beyond the scope of the book to cover this topic fully. My intention for this section is to provide students with a common starting point and to discuss some of the issues that can occur with the use of statistics in data analysis.

An important thing to establish before you use statistics is to understand why you are using it in the first place and what it is that you want to achieve. Statistics can be an important tool for various reasons, such as the need to (Graham, 2017):

- Summarise and represent data.
- Find patterns within a given data set, e.g. to establish if there are any significant differences between the different data sets.
- Find relationships, e.g. to see if two things correlate with each other and to further understand the strength of that correlation.
- Prove or disprove a hypothesis in a scientific investigation.

An important thing to appreciate is that there is a wide variety of statistical tools available out there. Although generally this is a good thing, it can actually pose a challenge to those who may not have a strong statistical background.

So, how do you go about choosing what is the best statistical tool to use for a given data set?

My advice to you is to understand how your scientific community uses statistics. As such, it is a good idea to invest some time reading a statistical book that has been specifically written for your particular scientific discipline, e.g. analytical chemistry (Miller & Miller, 2018), particle physics (Lista, 2016), biologists (Campbell & Richard, 1989) and engineers (Navidi, 2010). Another good way to help you identify the correct statistical tools is to take notes from similar studies that have been conducted in the past. It may be that you will need to know only the basics, such as the estimation of the mean, standard deviation, standard error, percentage, mode, median, variance, relative standard error %, etc. (Graham, 2017).

Statistical tools are important if you ever want to display your data, e.g. in a table or graph format. There are different graph types to choose from, e.g. bar chart, time graphs and pie graphs. In choosing what kind of graph is suitable, you must remember what your goal is, e.g. choosing a particular graph type to highlight a relationship within the data set. As such, you may want to draw a scatter plot and show a regression line (or the best-fit line) to subsequently report the correlation coefficient. In addition to basic plots/graphs, you may require the use of sophisticated statistical tools to display your data, such as the use of pattern recognition (Fukunaga, 1990) and machine learning (Bishop, 2006).

Overall, statistics have become more accessible to scientists as there are many suitable software packages on the market to choose from, with some common ones being Excel, MatLab, MiniTab, SAS, etc. Again, the kind of software that you will need will be dependent on your scientific discipline and the kind of study that you are conducting. Although statistical software is extremely useful and a huge time saver, choosing the right package can be a daunting process. Here are some of my tips on how to choose the best software for you:

- Seek expert advice from other scientists in your community with regard to their experiences.
- Always test the software before purchasing. Most software companies will allow you to have a free trial (usually for 30 days). During this time, make sure that you ask the vendor lots of questions (remember that vendors are usually keen to talk to you before you make a purchase). Particularly, you will need to understand what happens to your data when it is being processed.

- Do not start the 30-day trial without understanding the statistical concept behind what the software is trying to do and without having your own set of experimental data to test out.

When it comes to using any statistical software package, remember to:

- Always evaluate the quality of the raw data, to ensure that the data is indeed fit for its intended purpose, prior to statistical processing.
- Not to treat the software package merely as a black box. Understand how the software is processing your data.
- Identify any suitable pretreatment process. For example, you may want to remove unwanted sloping backgrounds in spectral fingerprints or conduct data normalisation prior to putting your data into a pattern recognition software. Ideally, you will want to come up with a pretreatment method that will treat all data set objectively and in the same manner. This may not be possible in all cases, e.g. if you need to reject outliers.
- Take care when presenting significant figures (SF) that represent precision. As such, do not merely copy and paste figures, without understanding the significance of those figures. For example, many students often quote a final figure with six decimal points, which not only represent excessive precision but the many digits can overcomplicate things and obscure the final message. So, what are the rules with regard to SF and how to round up numbers? There are no straightforward answers to this, as it all depends on your scientific discipline. For example, Cole has published a paper that summarises recommendations for rounding up certain common statistical values in medical scientific writing (Cole, 2015). From my own field of analytical chemistry, the general rule is to present all figures or digits in which you are certain of, and in addition quoting (and rounding off) one more figure after that, i.e. the first uncertain figure (McCormick, Roach, Chapman, & ACOL (Project), 1987; Miller & Miller, 2018). In addition, how you round up numbers will also depend on the type of study that you are conducting. For example, if you are taking part in a much bigger study (in which you have a role as a test laboratory), then you may not want to round off any figures until you have spoken to someone who will be ultimately responsible for the data analysis and to agree with him/her on how you are going to present the data.

6.3 Quantifying Measurement Uncertainty

What do I mean by measurement uncertainty?

The purpose of performing any measurement is to find out about a property of a particular item or object. At the end of a given measurement, you will end up with a value. Most likely, this figure is not the true value, as it will always have a degree of doubt associated. This degree of doubt is what we refer to as measurement uncertainty (Willink, 2013).

So, why is quoting measurement uncertainty so important?

The ability to quote measurement uncertainty allows you to have confidence in your measurement. This in turn will allow you to decide if that measurement is good enough for your particular needs. This can be vital for making important decisions, needed for example to validate a method, develop reference materials and verify/qualify products.

So, how does one quote measurement uncertainty?

In order to do this, you will need to perform what is known as an uncertainty budget. I will not cover this topic extensively here, as I will only give you a brief outline on what the process entails. Essentially, when you develop an uncertainty budget, there are two main steps associated, which involves the need to (Ermer & Miller, 2005):

1. Identify the various sources of uncertainty, i.e. quoting the different uncertainty components associated with the measurement. Each uncertainty component can be expressed as the standard deviation of that element, which in turn can be done by performing repeated measurements.
2. Statistically combine all of the different standard deviations from step 1.

If you want a step-by-step account on how to develop uncertainty budgets, then it is best that you refer to past references of relevance to your scientific area, e.g. chemical analysis (De Bièvre & Günzler, 2003), surface measurement in industry (Smith, 2002) and analysis of a dimensional inspection process (Los Alamos National Laboratory, 2012).

6.4 Achieving Measurement Traceability

You will need to decide on how accurate you want your measurement to be. If you want to have a high degree of accuracy in your measurements, then you will need to make the measurement traceable, i.e. the ability to trace back to something of a higher level. In the context of measurement traceability, this means that you are able to link your measurement to a higher state of measurement accuracy, achieved through an unbroken chain of measurement comparisons (for example, to a stated national or international reference) (De Bièvre & Güzler, 2005). Amongst other things, you will need to be able to perform instrument calibration and drawing up uncertainty budgets.

Achieving traceable measurement is not a trivial task. You would need to refer to case studies that are specific to your own scientific area, to appreciate how measurement traceability can be achieved, e.g. chemical (De Bièvre & Güzler, 2005), ionising radiation (Heaton, 1982) and temperature (Bentley, 1998).

Certainly, making traceable measurement is always desirable; however it may not be possible for all cases. This is particularly true for certain disciplines, e.g. biology and environmental science. With such scientific disciplines, you often get the analyte of interest residing in a complex matrix (e.g. biological or environmental).

As such, instrument calibration is often carried out under conditions too different from the actual application, which subsequently will make it impossible for you to realise traceability in your measurements (Conti, 2008).

6.5 Method Validation

Once a method is developed (as discussed in Sect. 6.1), it often needs to be validated for the purpose of uptake by the outside world to show that your proposed method is indeed fit for purpose. Hence, method validation is one stage after method development (Fajgelj & Ambrus, 2000). The steps involved in the method validation process can be quite extensive and the question on how to perform the validation will depend on your scientific field and circumstances. As such, if you are interested on how to validate your particular method, then you should refer to references on method validation specific to your particular discipline. For example there are detailed method validation approaches associated with the pharmaceutical analysis (Ermer, 2015), analytical methods (Chan, 2004), chromatographic methods (Bliesner, 2006), etc. Hence, goal for this section is to present some of the generic key steps often associated with method validation.

Key to method validation is the need for you to (Eskes & Whelan, 2016):

1. Plan your validation experiments carefully.
2. Write the relevant protocols. Such protocols are necessary so that you can tell other participants on what to do in a round robin study for example. As such, you will need to detail a step-by-step account on how to calibrate an instrument, qualify an operator, conduct the analysis, perform an uncertainty budget (if relevant), etc.
3. Carry out a small round robin study in the pre-validation stage, to quantify measurement errors and ultimately improve your method (Hester & Harrison, 2006). You can do this with a few established and competent laboratories to start off with. The purpose of the pre-validation stage is to allow you to refine your protocol, if necessary.
4. Carry out a formal validation trial. This usually involves a bigger round robin study, with other sponsor organisations. The difference between the pre-validation (that results in qualifying a method) and a formal validation stage (that results in the method being validated) is the depth and robustness of the two types of studies, e.g. the number of organisations participating in a given round robin study.
5. Record and develop the relevant documents. There are different types of documents that you will need to write and circulate, for the purpose of method validation. As stated above, when conducting a round robin study, you will need to provide participants with written procedures, detailing not only the step-by-step method but also their system suitability criteria prior to analysis. At the end of a round robin study, you will need to file a method validation report, detailing on

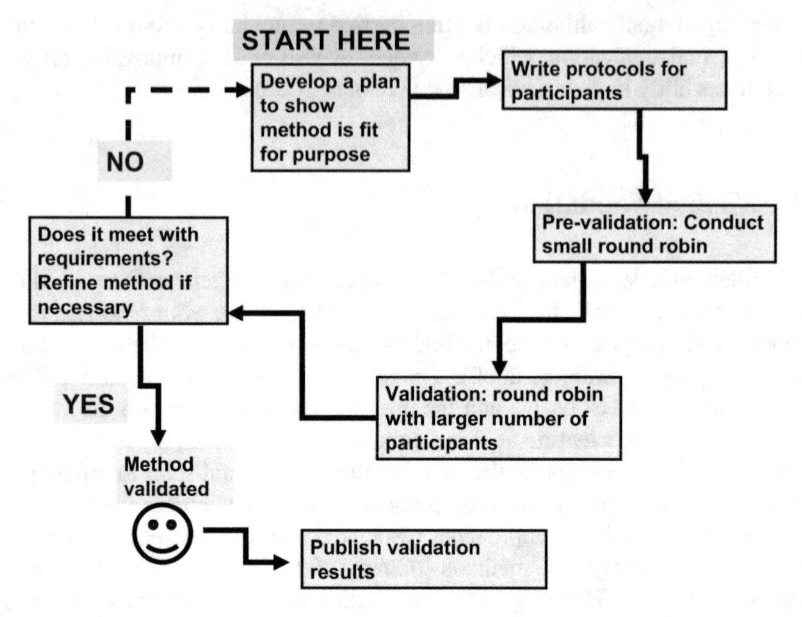

Fig. 6.2 The method validation cycle

how the study was conducted, and to also disclose final results to stakeholders (Snyder, Kirkland, & Glajch, 1997).

As with method development (Sect. 6.1), method validation may not be straightforward and you may find the need to refine the method (again and again) by going through the validation process more than once (as depicted in Fig. 6.2).

6.6 Laboratory Books: Best Practice on How to Keep Records

Primarily, a laboratory book is very much like a diary that details your investigations.

So, why is it important that you keep good records on what you have done?

A good laboratory book will not only benefit you but others, e.g. your colleagues, may want to extract vital information on how to do things, when you are not around. More importantly, a good laboratory book can act as a piece of evidence in the court of law, potentially when it concerns patent disputes (Graves & Graves, 2007).

So, what does a good laboratory notebook look like?

When you are recording information into your laboratory book, ensure that (Kanare, 1985):

- You write neatly and clearly (with a pen).
- Your laboratory book does not have removable pages.

- You date every page, to be subsequently signed by a senior member of staff (at the end of each day).
- You start on a new page when starting a new experiment. Remember to put a title and aim for each experiment at the top of the page.
- You write in sufficient details (on all procedures and observations), so that someone else can read and repeat the experiment.
- You cross out any mistakes (with a line going across), explaining next to it why it had been a mistake.
- You glue/tape securely any printouts (e.g. printed data, specification sheets).

6.7 Seeking Quality Assurance and Accreditation

Nowadays, in order to satisfy customers, companies from certain sectors may need to show that they have reached a certain level of quality assurance on, e.g., processes, data generation and laboratory practices. As evidence of this, companies may want to gain some kind of formal accreditation. There are various types of accreditation for quality assurance that can be sought for, to include good laboratory practice (GLP) (Special Programme for Research and Training in Tropical Diseases & World Health Organization, 2009), good manufacturing practice (GMP) (Nally, 2007) and good analytical practice (GAP) (Chan, Lam, & Zhang, 2011).

Before going any further, it is important for you to understand the difference between quality assurance and quality control. Let me illustrate this with an example. If you have a company that produces glucose sensors, then quality assurance is concerned with activities that are performed before the production of such sensors. Quality control on the other hand is concerned with activities performed AFTER the sensors have been produced. Hence, in quality control you evaluate the quality of the resulting product, e.g. if sensors have met certain analytical specifications (Konieczka & Namieśnik, 2009).

So, how can a company be accredited formally with regard to quality assurance?

In order to gain accreditation, the company will need to adhere to certain guidelines that are often set out in standard documents. These standard documents are often issued by globally recognisable standard organisations, such as ISO. Organisations will then need to participate voluntarily in external assessments such as the ISO accreditation programmes. Certification and registration will be provided by external companies that are qualified to make the assessment in the first place. In the UK for example, accreditation can be carried out by British Standards Institute (BSI) ("Standards, Training, Testing, Assessment and Certification | BSI Group", n.d.) or Bureau Veritas Quality International ("Certification, & Audit Services | Bureau Veritas Certification", n.d.).

There are several advantages as to why a company will want to acquire accreditation. Potentially, it could make the company more profitable by (Günzler, 1996; Seaver, 2003):

- Ensuring customer confidence, thus showing that the company is committed to maintaining a certain level of quality.
- Being more competitive in international markets.
- Complying with regulation: Some industries, e.g. the pharmaceutical industry, are required to have in place more stringent demands in relation to quality assurance.
- Encouraging and involving all employees to work towards a common goal: The workforce can potentially be more committed if this is the case, i.e. in which everyone will be expected to work towards an external, globally accepted standard.

Although the advantages of accreditation are clear, there are undoubtedly some disadvantages that you should be aware of (Günzler, 1996):

- The setting up of the system/infrastructure to achieve accreditation can be costly.
- The implementation step of establishing a quality system of a particular standard can be time consuming; for example the formalisation and documentation process may involve considerable secretarial expenses.
- The actual assessment and registration can be costly.

As such, laboratories often choose to acquire formal accreditation, if there is an actual requirement for one in the first place, e.g. when they need to meet certain trade and regulatory objectives.

6.8 Summary

As a research scientist, you must remember to incorporate and integrate the concept of best practice approaches in whatever you do, in order to increase your success in the laboratory. In this chapter, I have presented some common best practices that all experimental scientists should be familiar with, namely on how to develop method, analyse data using statistics, validate method, keep good records, improve measurement accuracy and gain accreditation for the purpose of quality assurance.

References

Ahuja, S., & Dong, M. W. (2005). *Handbook of pharmaceutical analysis by HPLC*. Cambridge: Elsevier Academic Press.
Bentley, R. E. (1998). *Handbook of temperature measurement*. Berlin: Springer.
Bishop, C. M. (2006). *Pattern recognition and machine learning*. Berlin: Springer.
Bliesner, D. M. (2006). *Validating chromatographic methods: A practical guide*. Hoboken: Wiley.
Bundesanstalt für Materialforschung und -prüfung (BAM) - RRR - COMAR Overview. (n.d.). Retrieved October 6, 2017, from https://rrr.bam.de/RRR/Navigation/EN/Reference-Materials/COMAR/comar.html.

Campbell, R. C. (Richard C. (1989). Statistics for biologists. Cambridge University Press Cambridge.

Certification & Audit Services | Bureau Veritas Certification. (n.d.). Retrieved October 6, 2017, from http://www.bureauveritas.co.uk/home/about-us/our-business/certification.

Chan, C. C. (2004). *Analytical method validation and instrument performance verification*. Hoboken: John Wiley & Sons.

Chan, C. C., Lam, H., & Zhang, X.-M. (2011). *Practical approaches to method validation and essential instrument qualification*. Hoboken: John Wiley & Sons.

Cole, T. J. (2015). Too many digits: the presentation of numerical data. *Archives of Disease in Childhood, 100*(7), 608–609. https://doi.org/10.1136/archdischild-2014-307149.

Conti, M. E. (2008). *Biological monitoring: Theory & applications—Bioindicators and biomarkers for environmental quality and human exposure assessment*. Southampton: WIT.

De Bièvre, P., & Günzler, H. (2003). *Measurement uncertainty in chemical analysis*. Berlin: Springer.

De Bièvre, P., & Güzler, H. (2005). *Traceability in chemical measurement*. Berlin: Springer.

Dulski, T. R. (1996). *A manual for the chemical analysis of metals*. West Conshohocken: ASTM.

Ermer, J. (2015). *Method validation in pharmaceutical analysis: A guide to best practice*. Weinheim: Wiley-VCH.

Ermer, J., & Miller, J. H. M. (2005). *Method validation in pharmaceutical analysis: A guide to best practice*. Weinheim: Wiley-VCH.

Eskes, C., & Whelan, M. (2016). *Validation of alternative methods for toxicity testing*.

Fajgelj, A., & Ambrus, A. (2000). *Principles and practices of method validation*. London: Royal Society of Chemistry.

Fukunaga, K. (1990). *Introduction to statistical pattern recognition*. Cambridge: Academic Press.

Graham, A. (2017). *Statistics: An introduction*. London: Teach Yourself.

Graves, H., & Graves, R. (2007). *A strategic guide to technical communication*. Peterborough: Broadview Press.

Günzler, H. (1996). *Accreditation and quality assurance in analytical chemistry*. Berlin: Springer.

Heaton, H. (1982). *Proceedings of a meeting on traceability for ionizing radiation measurements*. Washington, DC: U.S. Government Printing Office.

Hester, R. E., & Harrison, R. M. (2006). *Alternatives to animal testing*. London: Royal Society of Chemistry.

Hunt, L., Robitaille, D. E., & Williams, C. (2008). *The insiders' guide to ISO 9001:2008*. California: Paton Professional.

Kanare, H. M. (1985). *Writing the laboratory notebook*. Washington, DC: American Chemical Society.

Konieczka, P., & Namieśnik, J. (2009). *Quality assurance and quality control in the analytical chemical laboratory: A practical approach*. Boca Raton: CRC Press.

Kozak, A. (2008). *Introductory probability and statistics: Applications for forestry and natural sciences*. Wallingford: Cab International.

Lewis, I. R., & Edwards, H. G. M. (2001). *Handbook of Raman spectroscopy: From the research laboratory to the process line*. New York: Marcel Dekker.

Lista, L. (2016). *Statistical methods for data analysis in particle physics*. Berlin: Springer.

Los Alamos National Laboratory, U. S. (2012). *Uncertainty budget analysis for dimensional inspection processes*. US: Los Alamos National Laboratory.

McCormick, D., Roach, A., Chapman, N. B. (Norman B., & ACOL (Project). (1987). Measurement, statistics, and computation. ACOL by Wiley, London.

Miller, J. N., & Miller, J. C. (2018). *Statistics and chemometrics for analytical chemistry*. Carmel: Pearson Higher Education.

Murphy, C., & Yates, J. (2009). *The International Organization for Standardization (ISO): Global governance through voluntary consensus*. Abingdon: Routledge.

Myers, R. L. (2003). *The basics of chemistry*. Westport: Greenwood Press.

Nally, J. D. (2007). *Good manufacturing practices for pharmaceuticals*. London: Informa Healthcare.
National Research Council (U.S.). Committee on Reference Materials for Ocean Science. (2002). *Chemical reference materials: Setting the standards for ocean science*. Washington, DC: National Academies Press.
Navidi, W. C. (2010). *Principles of statistics for engineers and scientists*. New York: McGraw-Hill.
Osborne, J. W. (2008). *Best practices in quantitative methods*. Thousand Oaks: Sage Publications.
Seaver, M. (2003). *Gower handbook of quality management*. Farnham: Gower Publishing Ltd.
Smith, G. T. (2002). *Industrial metrology: Surfaces and roundness*. London: Springer.
Snyder, L. R., Kirkland, J. J., & Glajch, J. L. (1997). *Practical HPLC method development*. Hoboken: Wiley.
Special Programme for Research and Training in Tropical Diseases, & World Health Organization. (2009). *Quality practices for regulated non-clinical research and development*. Geneva: WHO on behalf of the Special Programme for Research and Training in Tropical Diseases.
Standards, Training, Testing, Assessment and Certification | BSI Group. (n.d.). Retrieved October 6, 2017, from https://www.bsigroup.com/en-GB/.
Tantra, R. (2016). *Nanomaterial characterization: An introduction*. Hoboken: Wiley.
Walker, R., Barwick, V., Bedson, P., Brookman, B., Burke, S., Lawn, R. E., et al. (2007). *Applications of reference materials in analytical chemistry*. London: Royal Society of Chemistry.
Willink, R. (2013). *Measurement uncertainty and probability*. Cambridge: Cambridge University Press.
Zhou, M. (2011). *Regulated bioanalytical laboratories technical and regulatory aspects from global perspectives*. Hoboken: Wiley.
Zschunke, A. (2013). *Reference materials in analytical chemistry: A guide for selection and use*. Berlin: Springer Science & Business Media.

Part III
Writing

Chapter 7
Peer-Review Publications

Let us imagine for one moment a worst-case scenario.

After months of hard work, you have made a wonderful discovery in the laboratory. Feeling happy, you enthusiastically report your findings to your boss or supervisor who also cannot contain his/her excitement. Without a second thought, he/she insists that you publish your findings. Without any guidance (and with no other mentors to turn to) you feel lost, as you have never done this before.

So, what do you do?

The purpose of this chapter is to answer this question, to help you publish your paper. As such, the first part discusses the publishing process, in which there are three main phases. The second part of the chapter aims to give you practical guidelines on how to write your paper. As there are different types of manuscripts (for peer review) that you can potentially write, it is not my intention to cover all types of articles, but to only focus on three common ones: technical, review and an opinion piece paper.

7.1 The Peer Review Process

Although different journals will have their own set way on doing things, the peer review process can generally be split into three main phases (as depicted in Fig. 7.1).

Phase 1 is concerned with the steps prior to manuscript submission. In this phase, you will need to:

1. Identify the type of article that you wish to write, e.g. technical, opinion and review.
2. Make a shortlist of journals that are potentially suitable. You should have some idea as to what journal you want to go for. This knowledge can come about from reading the literature or quite simply as result from having some kind of discussion with your supervisor.

R. Tantra, *A Survival Guide for Research Scientists*,
https://doi.org/10.1007/978-3-030-05435-9_7

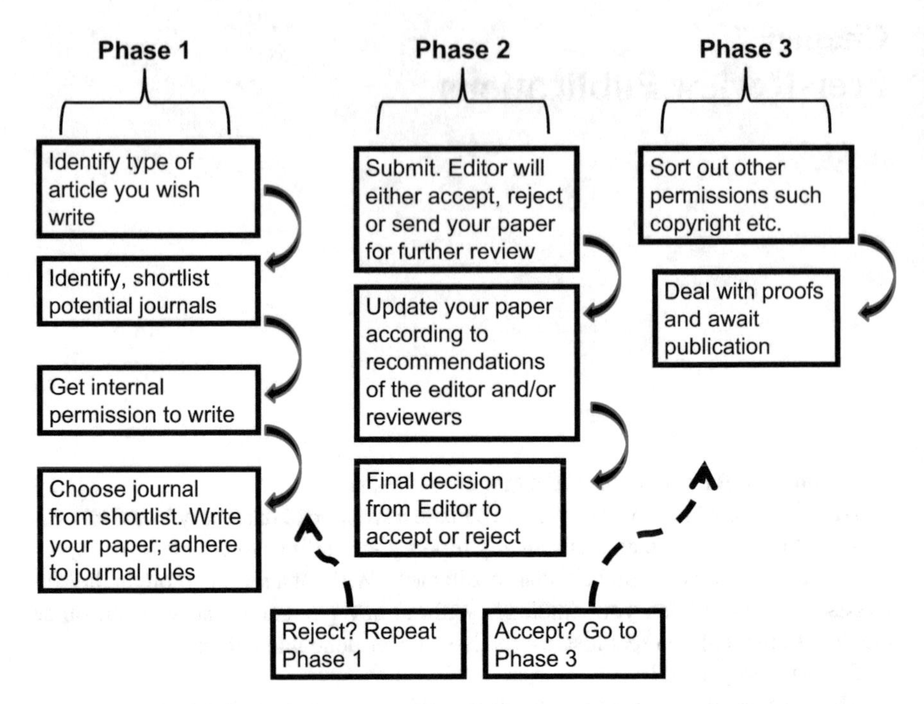

Fig. 7.1 The three phases to publishing your manuscript in peer review

3. Assess the suitability of each journal from the shortlist. Visit websites in order to identify the journal's requirements, as you will need to consider the likelihood of acceptance. You will need to take into account various factors, such as the journal's mission and impact factor (IF). Note: The journal's IF is an indication of its popularity and importance, which reflects on the yearly average number of citations to recent published articles (Mishra & Alok, 2017). Journals such as Nature or Science are highly prized due to their high IFs. However, getting something published in a high IF journal is not an easy task, as editors are looking for manuscripts that represent the frontier of innovation. Thus, your science needs to be groundbreaking and of substantial impact.
4. Choose a suitable journal from your shortlist.
5. Get permission from your boss (or supervisor) to gain his/her approval.
6. Write your article; please refer to the sections below on how to write different articles. When writing, please adhere to any rules imposed by the journal. Refer to the journal's website for guidelines that will advise you on structure of manuscript; style of writing, e.g. tense used; how to cite your references; word limit; how to display figures and tables, etc. When you begin writing, it is best to have in front of you examples of similar past papers that have been previously published in that journal. This is so you can identify and appreciate how things are written, as well as the level detail that you will need to include for your own paper.

7. Appreciate the timeline needed for you to complete phase 1. This will vary from one person to the next, as it will depend on how fast you write. As an approximation, it can take you up to a month to complete this phase. The process however can take longer, for example if you need to gain approval with other co-authors/consortium partners before submission.

Phase 2 (Fig. 7.1) happens after you have finished with your manuscript. At this stage, you will need to:

1. Submit your manuscript online, which will automatically go to an editor for initial assessment. He/she will make a decision of either accepting your paper straight away (from my experience, this is very rare!) or rejecting it on the spot, e.g. on the basis that your paper does not fulfil the original criteria set out by the journal. In most cases, the editor will send out your manuscript out for review, often to scientists with relevant background who will evaluate your paper on the basis of suitability, innovation, technical merit, credibility, science impact, etc. Although some journals can ask you to recommend potential reviewers, the final identify of your reviewers will always remain anonymous.
2. Wait for the results and then address any issues/comments that the editor or reviewers may have. Following the review stage of your manuscript, the reviewers will make their recommendations, to either reject or accept your manuscript. Usually, you are asked to further edit your paper or make comments to any queries that the reviewers may have. You will need to ensure that you tackle EVERY single query made. If certain comments seem unfair, then you will need to present a VERY strong argument to the editor as to why this might be the case. Once initial corrections have been made, you will need to resubmit your article. The editor then may want to send your manuscript back to the reviewers for their final consideration.
3. Wait for the final decision and further instructions from the editor.

You will need to appreciate that phase 2 can take some time, due to various reasons, such as the need to find suitable reviewers who are willing to do the job in the first place. Then you will need to allow several weeks (at least) for the reviewers to complete their assessment. Typically you can be waiting for a few months before you hear anything back from the editor.

Phase 3

If at the end of phase 2 your paper gets accepted then many congratulations, as you will now be entering into phase 3. In this phase, you will need to:

1. Sort out any copyright issues, e.g. sign a form to transfer copyright.
2. Go through any proofs that are sent to you, prior to the actual publication. It is vital that you read and check the accuracy of the final version before the manuscript gets published.

Please note that if your manuscript gets rejected at any time, then do not lose hope. This sometimes happens; it happened to me several times. However, it does mean that you will need to go back to square one again.

7.2 How to Write a Technical Paper

The key to writing a technical (experimental) based paper is to develop your story whilst you are still conducting your experiments. This will enable you to understand more about your process/system and will lead you to design experiments in order to fit a particular story line. It will allow you to stay focused in what you have to achieve, rather than just merely chugging away in the laboratory, churning out useless data. Remember that the more interesting your story becomes, the more likely it is for your manuscript to be accepted for publication.

After having some idea on what your story line will be, you will need to:

1. Develop an outline. This outline will highlight the different topics/sub-topics and the level of details that you will want to incorporate into your article (as illustrated below). The outline is important as it will clarify scope and will navigate you with what to write.

Title of Outline ...

1. Section 1: Introduction
   (a) Main topic 1 ...
      - Subtopic 1 ...
         – Details 1 ... (problem clarified)
         – Details 2 ... (arguments so far)
      - Subtopic 2
         – Details 1 ...

   (b) Main topic 2 ...
2. Section 2: Method
   (a) Main topic 1 ...
      - Subtopic 1 ...
         – Details 1 (Figure 1) ...

   (b) Main topic 2 ...
      - Subtopic 1 ...
         – Details 1 (Table 1) ...
      - Subtopic 2
         – Details 1 (Figure 2) ...
3. Section 3...4...5... etc.: Results ... Discussion ... Conclusion ... etc.

2. Collate all figures and tables that will support your story. The figures and tables should be able to *stand alone*. By this I mean that the reader should be able to look at the figures and tables to have a clear picture on what it is that you are trying to do, without having to read the body of text. When you are drawing up your figures and tables, remember to:

   - Have appropriate figure/table captions, to encapsulate the information that you are presenting.
   - Not duplicate information: For example, there is no need to do a table and a figure, if only one is needed in order to convey the same information.
   - Organise data/text correctly for a table, so that you are reading down a column rather than across the row (Kolin, 2015).
   - Choose appropriate figures to display your information (e.g. histograms, illustrations, drawings, photographs, flow charts, pie charts). For example, you can choose a scatter plot to show patterns or emerging trends between key variables (Budinski, 2001). Remember to plot your independent variable on the x-axis and your dependent variable on the y-axis. Also, remember to label both axes and include units of measurement. Avoid having too many plots in one graph; otherwise things can look cluttered.

3. Gain approval from your boss/supervisor. Show him/her your outline, collated figures, etc. and ask for feedback. Gaining approval early on is an effective way to ensure that your writing will not be a wasted effort.
4. Start writing. For more information on how to write and edit, please refer to Chaps. 10 and 11, respectively.
5. Come up with a title that will set the right tone for the reader and the journal that you are targeting. For example, if you want to publish for a high-impact paper such as Nature, then the title should be catchy. If you are going to submit your article to a journal with a more focused readership, then the title should be more descriptive, to entice the readership from the right circles (Lindsay, 2011). As such, you may want to include descriptive elements in your title, such as key parameter/s, the particular system (or aspect of the system) under study and the variable manipulated.
6. Write an introduction. The aim of the introduction is to give your reader background information. It aims to answer one major question: Why is your paper (or findings) of particular interest to the readers (or scientific community) which the journal is targeting? As such, you should (Beer, 2003):

   - Describe current knowledge or state of the arts.
   - Define and clarify the scientific problem or issue that you are trying to solve. Why are they important?
   - Define your objectives. What is it that you are trying to figure out? State any relevant hypothesis or predictions. Remember, a hypothesis is usually associated to an IF/THEN statement.
   - Briefly explain your method strategy. What will you be doing in order to achieve your set of objectives?

7. Write a method section. In this section, you will need to give a detailed, step-by-step account on what you did. You will need to give sufficient detail so that another researcher can duplicate your results in his/her own laboratory (Beer, 2003). Remember, you will be detailing not only procedures but also the materials that you have used. In addition, you will be including information on what: measurements were made, instruments were used, statistical tests/calculations were performed, etc. Overall, the style of your writing in the method section should be direct and accurate.
8. Write a results and discussion section. You will need to refer to the journal rules to deduce if you need to write this as one or two separate sections. For the sake of argument and to simplify matter, let us consider the situation in which you have to write it as two separate sections.

   In the result section, you will need to incorporate the appropriate text to accompany the figures and tables that you will be presenting in your article. The text in the results section should be a stand-alone piece. In other words, it should be of sufficient detail that the reader will not have to refer to the figures and tables, in order to follow the text. You will need to describe what you see and point out any key patterns/any interesting observations that arise from the data (Youdeowei, Stapleton, & Obubo, 2012). You do not need to describe everything that you see—you are not writing a diary. The results should detail any statistical tests that were carried out and overall encapsulate interesting/exciting findings to be discussed later (Meier, Zünd, Wiley, & Sons, 2000).

   In the discussion, you will need to expand on what you wrote in the result section. Although you can highlight significant results or key findings, do not repeat what you just said in the result section. Instead, you should comment and interpret what the results actually mean. As such, please bear in mind the following questions when you write the discussion section (Youdeowei et al., 2012):

   - Why are your results interesting?
   - How are the results linked to the original problem that you are trying to solve?
   - Does your results support your original hypothesis? If not, why not?
   - Did you get strange results? If your results are unexpected, you must explain the potential reasons why this might be the case.
   - Do your results compare well with past observations?
   - What other interesting scientific questions have arisen from your results?
   - What suitable future work are you proposing?

9. Write a conclusion. Here, you are summarising the key findings, highlighting the points that you would want the reader to remember and what these mean on a greater scheme of things.
10. Write an abstract; do this last. This typically consists of one (not more than two) paragraph(s) in length. The abstract gives a condensed version of the entire manuscript (Greenlaw, 2012) and contains:

- The problem that you are trying to figure out.
- The purpose of your study.
- What you did.
- What you found.
- What you concluded.

7.3 How to Write a Review Article

A review paper differs from a technical paper in that it does not report original research. It relies on existing literature, compares finding from different studies and is often written in order to summarise current state of knowledge. Potentially, it can focus to answer a particular scientific question or discuss a particular issue. Remember, some journals do not accept unsolicited review articles, so it is worthwhile to check if this is the case before you proceed.

When writing a review article, you will need to:

1. Draw up an outline, so that you can stay focused when you do write.
2. Ensure that you get approval for the outline, from your supervisor (and other co-authors).
3. Come up with a title. The title should be informative, concise and specific.
4. Write an introduction. This section will allow you to introduce your topic and gives sufficient background information to your readers. The aim of the introduction is to (Lindsay, 2011):
    - Underline the importance and significance of your chosen topic.
    - Present any current major lines of thoughts.
    - Present any conflicting views.
    - Clarify what the problem is and what you are planning to present or discuss.
    - Present your strategy on how you are going to answer the question that you have posed in the first place.
    - Discuss your methods briefly. Will you be relying on published literature? Will you be using grey paper materials such as government reports and project reports?
    - Discuss what will be presented in the main body of text.
5. Write a main body of text. Refer to your outline for guidance but do not be afraid to edit the original outline as you go along. Try to come up with accurate titles for the different sections and subsections in the main body of text. You will need to ensure that these headings/subheadings are arranged logically, so that ideas flow smoothly from one to the next (Lebrun, 2011). In essence, you will want to develop a pathway to consist of key thoughts that will eventually lead to your conclusion. You will not only need to present an up-to-date knowledge but you will need to make sense of it, i.e. what it all means in relation to the problem that you are trying to solve.

6. Write a summary. The aim of this section is to consolidate ideas and to remind readers of key findings, highlighting any interesting observations or patterns that emerged from your review (Altmann & Hallesky, 2011).
7. Write a conclusion. In this section, you will need to remind the reader of the significance of the topic, draw an overall conclusion to the study and give possible avenues for future research. Do not introduce any new data or results.
8. Write an abstract; do this last. Remember that the purpose of an abstract is to give a short summary of the entire paper of the problem that you are figuring out, what you did, the key findings and the overall conclusion.

For more information on how to write and edit, please refer to Chaps. 10 and 11, respectively.

7.4 How to Write an Opinion Piece Article

This type of paper should only be written once you have matured as a scientist and have established yourself in a particular scientific community. Your writing should have a certain level of maturity, to understand not only issues behind the science but also your audience being particularly sensitive to their concerns and what information they would like to know.

Writing an opinion piece in a journal is a lot more difficult than writing a technical paper or review paper. For a start, this type of paper does not follow any set format, even though the length of the paper is much shorter. The writing style must be clear. You will need to have a strong opening statement and an even stronger closing statement. In between, you will need to articulate different arguments clearly, such that these logically flow–from one point to the next.

So, how do you go about writing an opinion-based paper?

In order to do this, you will need to:

1. Decide on your *take-home* message.
2. Brainstorm to effectively identify main points that will support your take-home message. You can do this firstly by yourself and then with others. Whilst you are brainstorming, remember to make bullet points as you go along. From these, you will need to pick out a few strong points that you would like to include in your article (Laplante, 2012). Ask yourself:

   - Are these points strong enough?
   - Are there any points that repeat themselves?

   Edit your bullet points and then put them in order. Each bullet point will literally become a signpost that will allow your reader to logically follow from one point to another, eventually leading them to the take-home message (Jensen, 2006).
3. Think of a catchy title. This will not only attract your reader's attention but also set the tone for the entire article.

4. Write a body of text. Remember, an opinion piece does not follow a specific format and thus does not require the usual: abstract, introduction, results, etc. Usually a couple of paragraphs is all that is needed to make your point come across. For more information on how to write and edit, please refer to Chaps. 10 and 11, respectively.
5. Engage with your audience to make your story interesting. There are several ways in which you can do this (Rosenberg, 2005):
   - Clarify the problem at the start of the manuscript.
   - Come up with a strong opening argument.
   - Humanise your issues or concerns, i.e. make it real, so that you can connect with your audience better.
   - State key facts, with relevant references.
   - Communicate different theories and any alternative arguments. Show that you are not only thorough in your way of thinking but also thoughtful enough to recognise any opposing views. Remember that although you may be presenting opposing views, you should never fully agree with them.
   - Come up with a strong closing argument to identify the key points that you will want the reader to remember.

7.5 Summary

The main purpose of this chapter is to give you practical guidelines on how to publish in peer-review journals. To begin with, I have described what the publishing process entails, which essentially is made up of the three phases. Subsequently, I gave you step-by-step protocols on how best to write three very different types of articles: technical (experimental), review and opinion-based paper. The first two types are much easier to write, as they often follow a set format. Writing an opinion-based paper is a lot more challenging, though much shorter in length. Your writing style here will need to be a lot stronger, in order to present case/arguments well that will eventually lead to your take-home message.

References

Altmann, E. J., & Hallesky, G. J. (2011). *Technical writing that works*. Bloomington: Authorhouse.
Beer, D. F. (2003). *Writing and speaking in the technology professions: A practical guide*. New York: IEEE Press.
Budinski, K. G. (2001). *Engineers' guide to technical writing*. Ohio: ASM International.
Greenlaw, R. (2012). *Technical writing, presentation skills, and online communication: professional tools and insights*. Information Science Reference.
Jensen, J. N. (2006). *A user's guide to engineering*. Upper Saddle River: Pearson Prentice Hall.
Kolin, P. C. (2015). *Successful writing at work*. Boston: Cengage Learning.

Laplante, P. A. (2012). *Technical writing: A practical guide for engineers and scientists*. Boca Raton: CRC Press.

Lebrun, J.-L. (2011). *Scientific writing 2.0: A reader and writer's guide*. Singapore: World Scientific.

Lindsay, D. R. (2011). *Scientific writing: Thinking in words*. Clayton: CSIRO Pub.

Meier, P. C., Zünd, R. E., Wiley, J., & Sons, I. (2000). *Statistical methods in analytical chemistry*. Hoboken: Wiley.

Mishra, S. B., & Alok, S. (2017). *Handbook of research methodology: A compendium for scholars & researchers*. Bilaspur: Educreation Publishing.

Rosenberg, B. J. (2005). *Spring into technical writing for engineers and scientists*. Boston: Addison-Wesley.

Youdeowei, A., Stapleton, P., & Obubo, R. (2012). *Scientific writing for agricultural research scientists: A training reference manual*. Chicago: CTA.

Chapter 8
Reports

Throughout your career as a scientist, you are likely to write a number of different reports. The types of reports that you will be asked to write will depend very much on your personal circumstances, specifically the kind of organisation that you work for and the actual job that you do. Personally, I had noticed that when I moved away from academia and into government research (particularly when working on collaborative European projects), I had to write more and more reports, most being associated to reporting progress.

There are many reasons as to why you need to write a report. Reports can provide a record of past or current activities and thus be used as a way to co-ordinate/monitor projects. Reports can be used to communicate a technical solution to a problem or be used as a source of information in order to initiate future actions, e.g. making recommendations for business decisions (Souther & White, 1984). As there are a number of different possible reasons as to why you need to write a report in the first place, there are no set rules when it comes to format and style.

My goal for this chapter is to give you the necessary background knowledge, so that you can write reports with ease and thus divided into two main parts. The first part discusses the preliminary process of report writing, i.e. the thought processes that you will need to go through, before the actual writing stage. This is a useful starting, as by going through the correct initial thought processes, you will be able to develop a suitable structure/format for the report. The second part of the chapter will give you practical tips on how to write the report itself. Although I cannot tell you exactly what to write or what format to adopt, I will elaborate on some of the more common elements often found in reports.

R. Tantra, *A Survival Guide for Research Scientists*,
https://doi.org/10.1007/978-3-030-05435-9_8

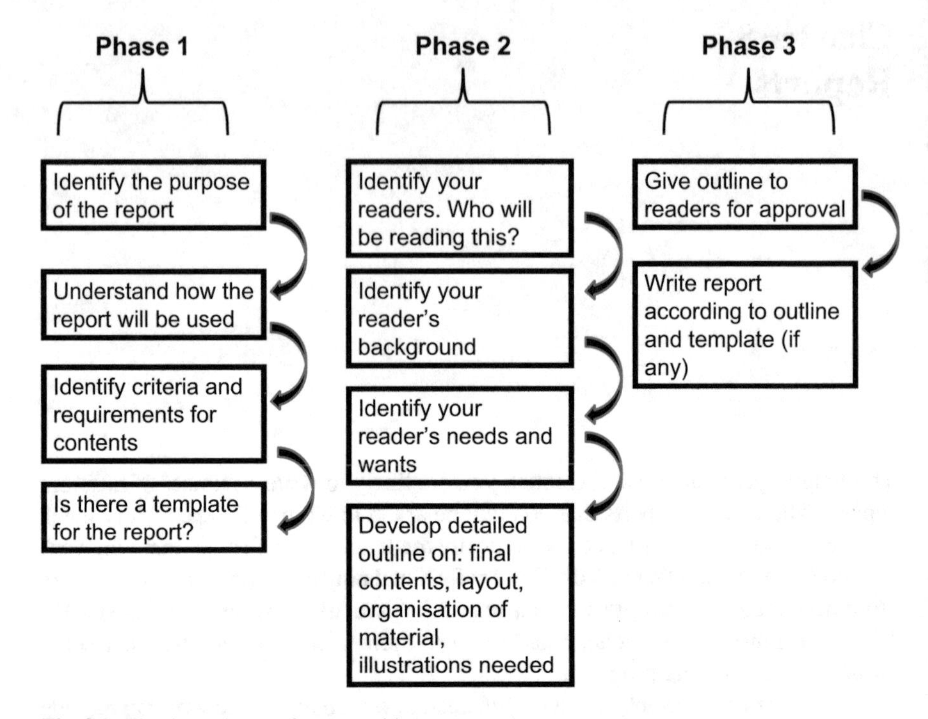

Fig. 8.1 The three phases of report writing

8.1 Preliminary Process of Report Writing

A key step in report writing is to go through the right thought processes in the preliminary phase, i.e. before you start to do the actual writing. This is important as it will help you clarify what material to include and identify an appropriate style/format for the report.

The thought processes can be divided into three phases (as depicted in Fig. 8.1):

1. Phase 1: In this phase, your goal is to clarify the purpose of the report (Riordan, 2005). Ask yourself:
   - Why have you been asked to write this report in the first place?
   - What will the report eventually be used for?
   - Have you been provided with a template?
2. Phase 2: Your aim in this phase is to understand your audience, more and to develop a detailed outline for the report in order to establish scope. Ask yourself (Souther & White, 1984):
   - Who will be reading my report?
   - What do my readers want to know?
   - What are their interests and background?

By answering such questions, you will be able to develop a detailed outline on what you want to write. This will help you to:

- Identify your pitch, in order to determine the kind of background information you will need to present.
- Clarify topics/sub-topics (plus any illustrations) that you will need to include.
- Organise the material, to determine a suitable layout that will ensure smooth flow of ideas. A sure way to create flow in your writing is to have good connections on how your text is organised, preferably establishing logical transition from one idea to the next (Souther & White, 1984).

3. Phase 3: In this phase, your goal is to gain approval from key people, e.g. those who have asked you to write the report in the first place. You can give your detailed outline (developed in phase 2) to your boss/supervisor. Getting feedback at this early stage is important to make sure that you have included all the material necessary for the report. Once an agreement has been reached, you can then finally write the report. Remember that when writing, you will need to refer to your detailed outline regularly, to ensure that your writing remains focused.

Note that the need to gain approval in phase 3 may not be necessary in every case single. For example, if your report is short and regular, e.g. doing a monthly report for your immediate boss, then I would skip this step.

8.2 What to Write?

Before proceeding any further, I would like to highlight that your report can be used as the basis of life-changing decisions (Kuiper, 2009). Therefore, remember to ensure that the presented written information is not only accurate but of sufficient details (preferably based on complete and thorough investigations). Your writing style should be clear for report writing, i.e. direct and to the point. It is not the time to use flowery or creative language! As previously pointed out, it will not be easy for me to tell you exactly how to write your report, as the format/style/content/layout, etc. can vary from one report to the next. Thus, the aim of this section is to discuss some of the common elements that are often found in most reports.

8.2.1 Title of Report

You will need to come up with a title that will accurately and clearly describe the purpose of the report.

Historic timeline of report

Date	Description	Filename
15/01/2017	RT creates report outline and sends off file to JT, HH for approval	V1_Outline.doc
17/01/2017	RT receives feedback from JT and HH.	V2_Outline.doc
20/01/2017	RT sends updated outline to JT, HH for approval	V3_Outline.doc
21/01/2017	JT, HH approves outline	V4_Outline.doc
01/02/2017	RT creates first draft of report and sends out to consortium partners for feedback	V1_Report.doc
... ETC.		

Fig. 8.2 History of a report: an example

8.2.2 History of Report

This part can be included if you want to show your readers the historic timeline of how your report has developed through time, for the purpose of transparency. As exemplified in Fig. 8.2, you can start from a time when you have created the report outline. As the report changes and evolves through time, you will need to record details on relevant activities, e.g. who has read/edited the report, when it was read, associated filenames, etc.

8.2.3 Dissemination Level

The can be included to inform your readers if the report is to be made public, confidential or restricted to a particular group of people.

8.2.4 Table of Contents

This will be necessary for long reports. You should list heading and subheadings, to appear exactly as in the text of your report and display the associated page numbers (that these commence).

8.2.5 *Abstract or Executive Summary*

This is an overview of the entire report, often placed at the front where readers cannot miss it. Having an abstract is useful if your report is particularly long and complex. Usually, this section is written for readers who may be busy and does not have time to read the entire report.

Please note that the word abstract and executive summary are often interchangeable nowadays, as it often means the same thing. However, to be more accurate and if a slight distinction is to be made, then the two usually differs in length. An executive summary is usually between two and four pages, whereas an abstract is often a paragraph (or two max.) in length.

By giving a short condensation of the report content through the abstract, you will be giving your readers (Weatherford, 2016):

1. A brief description of the problem.
2. An overview on the approach used to solve the problem.
3. Key highlights associated with results/findings.
4. A short conclusion, to include any key recommendations, e.g. for the purpose of management decisions.

8.2.6 *Introduction*

The aim of the introduction is to give your readers background information, so that they are able to follow what you have written in the body of text. It is best to avoid writing long and elaborate introductions for reports, as you will want to include only essential information that is considered to be essential (Aidoo, 2009). For example, if you are writing a report on product development, then your introduction should state what was wrong with the last product version and how the new version will benefit the customer. If your report is on literature review, then the introduction should clarify the problem/issue that you are trying to figure out, why it is important, the scope of the review and the audience in which the report intends to target. State what you know so far, particularly in relation to other similar works, state-of-the arts/updated technical knowledge, fundamentals/theories, etc.

Overall, when writing your introduction, make sure you (Aidoo, 2009):

1. Know your audience.
2. Define the problem clearly.
3. Identify the objectives, stating why you are writing the report in the first place. What you are trying to find out? What do you expect to achieve? What questions are you set to answer?
4. Include essential background information.
5. Underline the scope of the report, e.g. what will (or will not) be included in your report.

8.2.7 *Method*

In the method section you will need to detail on how you did work, such that you are able to reach your stated objectives. For example, if you have carried out a survey as part of your method, then you will need to say something about how you chose your subjects, how you came up with the questions to put in a questionnaire, how you analysed the survey results, etc.

When writing the method section, you will need to consider the following questions (Kmiec & Longo, 2017):

1. How did you determine the approach?
2. What did you do?
3. What equipment did you use (if any)?
4. How did you gather the data?
5. Were there problems that you had to overcome?
6. How did you evaluate the data? Did you use any statistical treatment or tests? If so what?

8.2.8 *Results and Discussion*

For this part, you can have the results and discussion as one section or as two separate sections. For the sake of simplicity, let us pretend that you have decided to write these as two separate sections.

In the result section, you should present your findings as plainly as possible, without making any comment or interpretation (Souther & White, 1984). You will need to summarise your key findings and display them in figures or tables. Decide early on how much data you will want to put in this section. Your aim is to put sufficient data, in order to satisfy the initial objectives stated in your introduction.

In the discussion section, you will need to assess the data presented and discuss what these mean. You can do this by presenting the reader with your interpretation of the data and apply associated reasoning (Souther & White, 1984). Overall, the discussion section will allow you to:

1. Compare the results with other findings, e.g. historic or published data.
2. Compare your results to any expectations or predictions that you might have mentioned in your introduction.
3. Identify any anomalies or strange results (in which you will need to provide an explanation for).
4. Highlight and discuss any significant findings.

8.2.9 Summary of Activities or Progress

This part is often needed in order to monitor the progress of a project e.g. when reporting to management. You should include details on what you have done during a particular timeframe, for example:

1. What you have achieved.
2. How outputs have been disseminated/exploited.
3. What impact or benefits there might be to stakeholders (e.g. public) as a result of the work that you did. Has the work had led to new collaborations or funding streams?

8.2.10 Conclusion

The conclusion should follow on logically from the body of the report and thus be presented last. You will need to highlight key findings from the study and make recommendations, e.g. future actions (Riordan, 2005).

8.2.11 The Use of Topic and Sub-topic Sentences for Headings and Subheadings

Although including sections such as the introduction, method, results and discussion, and conclusion may be suitable for some reports, e.g. to report the outcome of an experimental investigation, it may not be necessarily suitable for others. Hence, you might like to consider the use of topics or sub-topics instead, for headings and subheadings, respectively. These headings and subheadings will thus form the main structure in the body of text. It is important that you must reveal a clear organisational relationship between them, so that there is a logical flow of ideas throughout (Riordan, 2005).

8.2.12 Appendix

The appendix can be classified as the non-essential part of the report. This section can be useful in order to provide further explanation to the body of the report, e.g. to further support your analysis (Riordan, 2005).

8.3 Summary

My goal for this chapter is to present a logical approach, on how best to tackle report writing. It has not been easy for me to give you a definite guide on what to do when it comes to report writing, as the structure and format will ultimately vary from one report to the next. Ultimately it will be governed by its purpose, i.e. why you have been asked to write the report in the first place.

An important aspect of report writing is to go through the initial stages, i.e. the thought processes that you will need to go through, before you enter the actual writing phase. This preliminary stage requires the need for you to understand your audience, to clearly identify their needs. Subsequently this will allow you to develop a suitable outline that specifies content and scope. When it comes to the actual writing of your report, I have presented some practical tips on what sections you can potentially include.

References

Aidoo, J. (2009). *Effective technical writing and publication techniques: A guide for technical writers, engineers and technical communicators*. Leicester: Matador.

Kmiec, D., & Longo, B. (2017). *The IEEE guide to writing in the engineering and technical fields*. Hoboken: John Wiley & Sons.

Kuiper, S. (2009). *Contemporary business report writing*. South-Western Cengage Learning.

Riordan, D. G. (2005). *Technical report writing today*. Boston: Houghton Mifflin Company.

Souther, J. W., & White, M. L. (1984). *Technical report writing*. Huntington: Krieger Publishing Company.

Weatherford, D.-J. (2016). *Technical writing for engineering professionals*. Tulsa: PennWell Corporation.

Chapter 9
Bid Proposals

Let's face it: doing science is not cheap. As such, there comes a time in your research career that you will be asked to get involved with bidding, in order to keep your research alive. Quite often, you will not be working on your own, as the development and submission of a bid proposal is often a team effort. Whatever your role may be in the bidding process, it is important that you have sufficient knowledge to be able to deliver your part with confidence. The goal of this chapter is to give you that knowledge. As there are different types of bids, it would be impossible for me to cover this topic extensively in a single chapter. As such, I have decided to focus only one type of bid proposal, i.e. research grants, which often generates the largest funding streams (particularly for academic organisations).

This chapter is divided into three parts. The first part gives you background information, detailing the grant process (in which there are three main phases). The second part of the chapter discusses some of the common elements often found in grant proposals. I will also give you tips on how best to write these elements. The third part of the chapter gives you an insight as to why some proposals end up getting rejected. By knowing the potential reasons for failure, you can avoid common mistakes, in order to improve your rate of success.

9.1 The Bidding Process

Let us take a look at a likely scenario.

You come across a bid call and read the accompanying text. You get excited. It is very much up your alley! You tell your boss, who urges you to put something in.

After all it won't hurt?

… or will it?

Certainly, it will not hurt your boss, but remember that it may hurt you, in that you will be dedicating at least several weeks of your time (if not more) to develop a suitable proposal for submission. You must understand that securing funding is not

R. Tantra, *A Survival Guide for Research Scientists*,
https://doi.org/10.1007/978-3-030-05435-9_9

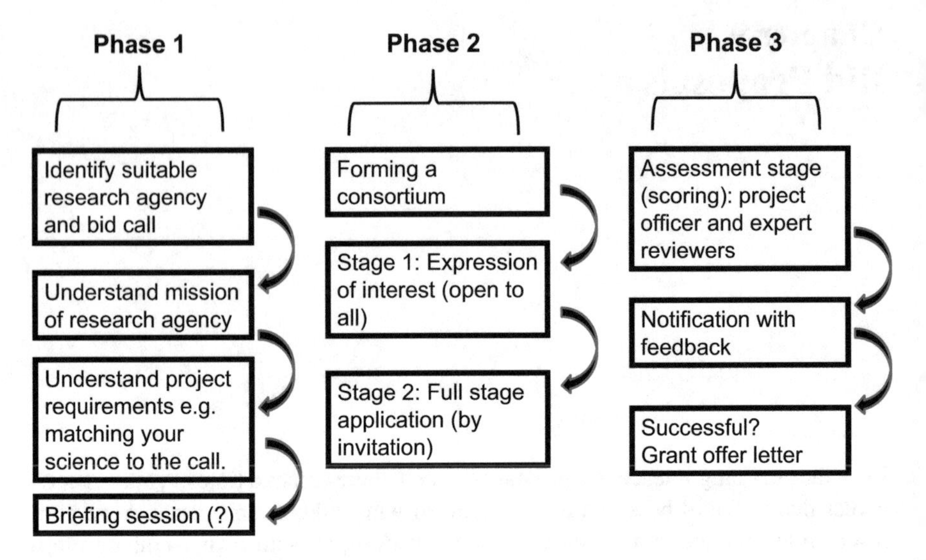

Fig. 9.1 Bid (grant) process: the three phases

a trivial task. It can be extremely hard work and very time consuming. If you are not careful, you can waste a lot of time *chasing a red herring*.

Hence, in order to ensure that your efforts will not be in vain, you will first need to appreciate just how the bidding process works.

Figure 9.1 depicts what lies behind the bidding process, with specific focus on how to secure funding for a scientific grant proposal. The process can be divided into three phases:

1. Phase 1: In this phase, your goal is to identify a suitable research agency and an appropriate call that will support your science. You can do this the hard way, e.g. by trawling through the Internet, or you can make it easy for yourself, by relying on your contacts/network (who will also be in the same position as you are and who may already have the vital information at hand).

   Once you have found a suitable agency and the corresponding call, you will need to understand the needs of the agency. Ask yourself:

   - What is the mission of the research agency? What is important to them? What is their programme of interest?
   - Is the call geared towards method development or groundbreaking innovation? What is the expected technology readiness level (i.e. maturity of technology) at the end of the project?

   Following on from this, you will need to identify the challenges and identify if there are any good reasons as to why you cannot join to bid. Remember, it is better to kill the proposal early on, rather than wasting further efforts only to later find out that you were not able to participate in the first place.

Ask yourself:

- Is there a good match between the science that I am doing and the requirements of the call?
- Who should be leading the bid? Will I need to find collaborators? If so, which organisation/s will I need to work with?
- What is the bid timeline? Realistically, can I commit to this?
- Are there any budget constraints, such as limitations on travelling expenses? Can I purchase capital equipment?
- Is my organisation eligible?
- Will my organisation and boss support the idea?
- Do I need co-funding? If so, will I be able to access this?
- Do I foresee any intellectual property (IP) issues?
- What is the expected funding success rate?
- Etc.

When collecting preliminary information during phase 1, remember to jot down any concerns and questions that you might have. If you are not able to find the answers to your questions, e.g. from the agency's website, then there will always be someone that you can contact, either a national contact point or the agency's project officer. In deciding whether or not you should proceed forward, remember to use common sense. For example, if you are still an inexperienced scientist, then you should not aim to lead/co-ordinate a proposal, particularly if the success rate associated with funding approval is poor (e.g. less than 25% acceptance rate).

Quite often, research agencies will schedule some kind of a briefing session open to all potential participants. If at this stage you have established your suitability (and you are keen to be involved in the bidding), then you must attend the relevant briefing session. Not only will you be able to understand more about the call but you can also talk to like-minded people who will also be interested in participating. Such meetings are generally useful, as it will give you a chance to not only maintain but also extend your network contacts. At such events, you can introduce yourself to project officers and other potential project partners. Who knows, you might be lucky enough to get invited to join a consortium that already exists!

2. Phase 2: In this phase, your goal is to join a consortium (or help a coordinator to form a team). When figuring out on how you can do this, it is worthwhile to do some background research, by talking to your network, to find out more about the competition. Who are the key players that are likely to bid? Is there an established consortium that you can join? Remember, joining an established (and strong) consortium will save you a lot of time/efforts and thus will increase your overall success.

   Once a consortium has been established, consortium members can help the coordinator to develop a suitable bid proposal for subsequent submission. The submission process can vary from one research call to the next. For example, it may be that you need to submit only one proposal. On the other hand, the submission may consist of two separate stages:

- Stage 1, in which you submit a short version of your project ideas: Subsequently, this will be used by the research agency to filter, to decide which consortium to invite for the submission of a full proposal (in stage 2).
- Stage 2: This is when you will need to develop and submit the full proposal. This stage is a massive effort and consortium partners often meet face to face to discuss work details, establish consensus and agree on the contents prior to its submission.

3. Phase 3: This is when all proposals have been submitted and the agency then enters the evaluation phase. All proposals submitted will first be reviewed by the project officer in charge, to assess initial suitability. Afterwards, these are sent out to technical experts who will judge the proposal further. The scientific experts are often chosen because of their expertise knowledge. They will usually judge the proposals against three main criteria: science excellence, science impact and project implementation. Proposals are often judged against a scoring system. It is often the case that proposals must reach a certain level of threshold if it is to be considered further. Proposals that have reached the necessary threshold level will then be ranked against one another. Projects are usually approved as a result of their ranking scores versus budget availability.

If you are one of the lucky few whose proposal has been accepted, then the project officer will offer you a grant letter informing you of this whilst reminding you of the conditions of the grant. If your proposal has been rejected, do not get disheartened. Do not throw away the feedback comments, stating why your proposal has been rejected. Learn from your mistakes and quickly move on to the next one.

9.2 Common Elements in a Proposal

The aim of this section is to give you an insight into some of the elements that are commonly found in grant proposals. However, it is imperative that you adhere to any rules that have been set out by the research agency. The research agency will make it clear to you as to what forms need filling and the information that is required. They will also dictate other particulars, such word limits and bid deadlines.

9.2.1 Table of Contents

Placed at the beginning of the proposal, this lists the different headings and subheadings in order of their appearance, alongside with corresponding page numbers (from start to end).

9.2.2 Title and Abstract

The title and abstract should be accurate, informative and able to *stand alone*. In other words, a reader should be able to read both the title and the abstract, to deduce what the project is all about without having to read the rest of the text.

The title needs to be the right length (i.e. not too long or too short) and informative, i.e. enough information to know what the document is about.

The abstract should only be a paragraph or two in length. The purpose of the abstract is to give an overview or a summary of the entire proposal, to detail (Frederick, 2011):

- What the problem is and why it is important.
- What you are going to do.
- How your research will change things.

Ensure that you match or align ideas in relation to the nature of the bid call. So, if the call specifically wants cutting-edge science to solve an ongoing problem, then you will need to highlight how your research idea will meet this challenge.

If you happen to lead a bid proposal, it is helpful to write a draft abstract, so that you can send it to potential project partners, so that they will know more about your research idea prior to commitment. It is often the case that after discussing the project idea further with other consortium partners, the initial idea can evolve and change. This is sometimes necessary in order to improve the overall quality of the bid proposal. As such, remember that the abstract gets updated just prior to submission.

9.2.3 Participant Information

This section gives a summary list of the partners in the consortium, to include any involvement from third-party organisations who may not necessarily benefit financially from the proposal. The bid coordinator will need to collate essential information for each participant, such as the names of key staff who will be directly involved in the project, their relevant area of expertise and activities, details of any relevant publications and organisational setting. The coordinator will need to convince the research agency that all participants are credible, e.g. individuals concerned have the right background experience to do the job and the organisation that they represent has the right technical infrastructure to support project delivery.

9.2.4 Introduction

The aim of the introduction is to give essential background information. This should be enough so that readers can understand the document and to enable the creation of a visual picture for the reader that is appealing. There are several important elements that you should consider when writing the introduction, namely the need to (Blackburn, 2003):

- Define the problem clearly.
- Clarify the concept behind the research idea. As such you may want to include historical information, theoretical background, etc.
- Clearly state the drive behind the research idea. Why is there a need to fund the research idea? What will happen if it does not get funded? What useful findings will the idea generate?
- Clarify what you are proposing to do, in order to solve the problem. What are your main objectives? Roughly state what you will be delivering (but do not go into huge details, as this will be covered elsewhere in the document).
- Define your competitive advantage. How is the project idea different to what has been done in the past?
- Highlight how the research proposal matches well to the requirements of the call text.
- Highlight any technology proposed and identify the technology readiness level, i.e. level of maturity (if appropriate).
- Have a clear goal beyond the lifetime of the project. Will your research idea result in a commercial product? Can the project outcome be used by other stakeholders, e.g. regulatory bodies?
- Clarify the *bigger picture* on how your research idea fits into the grand scheme of things. For example, how does your project ideas link with external projects or initiatives?

9.2.5 Methodology or Approach

This section is all about detailing the approach, i.e. what needs to be done in order to fulfil your stated objectives. It thus answers the questions of what (are you going to do), where, when, how and for how long. Your method should be logical, realistic (not overly ambitious) and consistent. You will need to have a clear vision of the expected results and identify any issues that you may encounter along the way.

You will need to detail the work plan proposed, which must be clear and well structured. It is useful to show flowcharts that sketch the different components of your overall work plan. You will also need to show links between different (clusters of) tasks/activities (Lewis, 2007). In a complex project, it is common to display a macro work plan first before breaking the major components into separate subcomponents in a micro work plan. In the flowchart, you can highlight any important points, e.g. where targets need to be met or if there are breakpoints, i.e. trends or relationships that should be analysed/discussed with partners before proceeding further.

In the methodology section, it is important that you identify any red flags, i.e. major issues to the project. For example, these can include:

- Issues related with data quality, e.g. how data are to be collected, analysed and handled.
- If you foresee any negative results that could impact the successful completion of the project, such as technical difficulties with a new technology.
- Etc.

You will need to discuss briefly how these problems can be minimised or be removed altogether. If you do not flag up such issues in your proposals and these get picked up by the reviewers, then this will heavily dampen your science credibility.

9.2.6 *Ethical Consideration*

You will need to declare any ethical considerations associated with your project, for example if you intend to carry out animal or human experimentation/s (Foster-Gilbert, 2001; Hubrecht, 2014) or if you intend to use extremely toxic chemicals. Under this umbrella, you can also detail on how you plan to interact with minority groups. For example, you may want to propose a strategy to ensure that enough women are being employed, in order to achieve a better gender balance.

Finally, do remember that it is not ethical to ask for money from different agencies to do the same type of work. It is imperative for you to declare any similar activities already funded or to state any overlapping work proposals.

9.2.7 *Impact*

All research grants will always ask you to write something on impact. In general, impact is about answering the question:

Will your research idea result in anything useful?

Remember, there is not much point in funding the project if the final outcome is not of use to anyone (no matter how excellent or innovative the proposed science is).

Please bear in mind that the kind of impact that you should write will be governed by the nature of the bid call and the mission of the research agency. For example, if the mission of the call is to develop a training infrastructure for young scientists in a specific country/region, then your impact section should be focused on highlighting how your research idea will enhance the skills and development of these scientists, to ultimately ensure that the young scientists trained will have jobs at the end.

You will need to detail on how impact will be realised. As such, you will need to write about:

- The dissemination route: Here, you will need to talk about how your findings will eventually get transferred to the intended target audience, i.e. those who will benefit from knowing the details of the research outcome. As a result, you will need to clarify suitable dissemination routes, e.g. publishing in high-impact journals, organising a dedicated conference and organising training events.
- The exploitation route: This details on how you can further exploit the proposed project outputs. For example, if you foresee innovation, your exploitation route can involve the filing of an intellectual property and subsequently asking local government to fund and help initiate the formation of a spin-out company that will one day create jobs for hundreds of workers.

9.2.8 *Project Implementation: Management and Project Delivery*

All bids will contain a section on how the project will be managed, to ensure that the project will be delivered successfully (within budget and on time) (Lewis, 2007).

In forming a management team, there is a need to choose carefully, in relation to the individuals who will eventually become team leaders, work package leaders, task leaders, intellectual property (IP) manager, etc. These individuals are key to the success of project delivery and preferably should have the necessary past experience

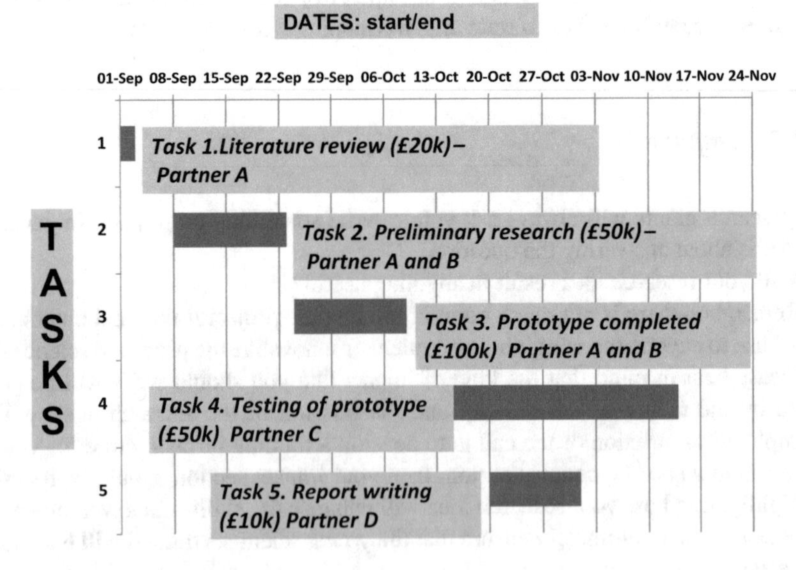

Fig. 9.2 Example of Gantt chart

to efficiently co-ordinate day-to-day work. For example, for the role of a team leader, it is wise to choose someone who has successfully led teams or consortiums before. Whatever management structure you choose, there should be a clear line of accountabilities with appropriate procedures in place, e.g. how management will handle future disagreements such as IP disputes.

A pictorial level diagram is useful to describe management structure and show hierarchy. It can also show how management will communicate, not only with the project partners, but also with other interface groups such as technical committees or steering groups (who will be overseeing the work at a more strategic level).

In order to convince the reviewers and project officers that your project is both cost and time effective, you will need to incorporate appropriate visual aids. For example, a management software such as Microsoft Project will help you to form charts, such as a Gantt chart, to show readers when a certain task will finish/end. Figure 9.2 illustrates an example of a Gantt chart.

A Gantt chart is often necessary, to show resource commitment and to justify costs associated with individual participants. It will clarify who is responsible for what, how much time they will spend, if any sub-contracting will be necessary, etc. Overall, you will need to ensure that the work proposed is well balanced to support the overall objectives of the project. You will need to include the appropriate supporting text (i.e. details on any planned activities) when describing your Gantt chart. For further clarification much of the information that supports the Gantt chart can be tabulated in a matrix-type format. A matrix table is useful to summarise:

- Activities associated with the individual work packages (WP)s and tasks: Remember to give WPs, tasks, etc. reference numbers.
- A list of milestone of targets and project outputs/deliverables.
- A list of foreseeable meetings (e.g. kick-off meeting, general assembly).
- A list of identified risks to the project and proposed mitigation actions.
- Etc.

9.2.9 Supporting Information (e.g. Budgets, CVs)

With regard to adding supporting or supplementary material into your proposal, it is vital that you adhere to the rules set out by the research agency. Do not add on any unnecessary material if these have not been specifically asked for by the funding body, such as printouts of papers that you have published. By all means add essential sections such as glossary and abbreviations.

If you are a coordinator then it is likely that you will need to collate information on project partners, e.g. CVs and budget requirements. When collating supporting information from partners, you may want to have templates readily available to give to participants. Such templates will help to ensure consistency, in relation to the content reported by the different partners, thereby allowing you to build an image of integration with ease. When submitting CVs of key individuals, you will need to include essential information such as (Nickson & BCS, 2012):

- Personal data, e.g. name, nationality.
- Background education, professional status and key qualifications.
- Field of specialisation and relevant experience: These can be set out as bullet points (in reverse chronological order). Don't go overboard with the information; 4–5 bullet points is more than ample for this purpose.

When giving details on budget information, it is necessary to justify the money that you and the rest of the project partners are asking for, from the research agency. You will need to have an idea on how much you can ask for, thus determining how ambitious your project should be. You can look into past funded proposals, to deduce how much budget has been allocated for similar activities; it is sometimes possible to get this information from the research agency's website. Please be aware that the budget that you are asking may not be the actual budget that you will get. So, when preparing budgets, you will need to have a plan already in mind on how to scale back, in case you are asked to quickly shorten the project or cut off a certain percentage from the total budget.

In developing the budget information, you will need to collate essential information from participants, to include (Cleden, 2011):

- Salaries, for those key members of staff who will have direct involvement in the project.
- Capital equipment needed.
- Indirect costs, e.g. overheads: Overheads are associated with those costs that cannot be attributed to a particular project, e.g. rental space, electricity bills, support staff (such as technicians) and equipment servicing.
- Research expenses, e.g. consumables (reagents, glassware, etc.).
- Travel and subsistence expenses, e.g. to attend meetings and scientific conferences.
- Co-funding requirements: Some research agencies require that participants match the funding from their institution or national body/organisation. You will need to verify that your organisation will be able to secure the required amount.
- Any subcontractors you plan to employ.

In preparing for your own budgeting information for submission (to the project coordinator), it is important to go over this information with someone in your organisation who has more experience, e.g. an in-house project finance officer. Furthermore, you can ask the finance officer if there is already a template ready for you to fill, e.g. in the form of an Excel spreadsheet that details budgeting information specifically for bid proposals. When doing your own budget justifications, remember to adhere to the rules set out by the research agency in relation to certain costs, e.g. spending restrictions on travel, consumables, capital equipment, etc.

9.2.10 General Information: Graphical Information

Whenever possible, you should try to use powerful graphics as a way to share your project vision with the reviewers and to achieve a more professional look. You can do this with the aid of commercial software such as CorelDraw, Visio and MindMapper. Choose the right type of graphics as a way to strengthen your particular message or idea. There are different types of graphics to choose from, to include network diagrams, flowcharts, Gantt charts, work tree, tables, matrices, etc.

When doing graphics, remember to (Lewis, 2007):

- Include a reference number, along with the appropriate figure legend.
- Have accompanying text to further clarify the graphic (for example, if a figure is to contain shades of colour/s that represents something or if a figure requires a scale bar).
- Choose suitable layouts, e.g. page setup.
- Avoid cramming too much information into graphs.
- Label axes on graphs clearly.
- Indicate the origin of the graphic information, if taken from another source.

9.3 Things to Avoid: Reasons for Rejection

It is important that you keep any feedback from your proposals that have been rejected, so that you learn and avoid making the same mistakes in the future. Below is a list of some of the common reasons as to why your proposal can be downgraded. By understanding some of the reasons for rejection, I hope that you will avoid making such mistakes. For ease of reading, I have grouped them into six separate categories.

9.3.1 Not Fulfilling Essential Criteria (Cleden, 2011)

- Participant/s not eligible for funding.
- Proposed idea not in scope, e.g. does not sufficiently address the issues highlighted in the call text. Hence, you are targeting the wrong call or even the wrong agency.
- Missing section/s in the provided form.
- Not abiding to basic rules, such as word limits and deadlines.
- Unanswered safety and ethical questions, e.g. no procedures disclosed for animal or human experimentations.
- In the case when matching funding is mandatory, there is no clear indication of funding from other sources.

9.3.2 Science Is Weak (Peach, 2017)

- Science credibility is questionable, e.g. technically not sound. As such, it is always worthwhile running past your idea to another expert in the field.
- Are you proposing to use an out-of-date technology? Why have you not considered better ones that have emerged since? Are you proposing a piece of science that has already been done before? Is the approach not logical, e.g. such as correlating two things together that are not directly linked? Does the proposed science sounds like a fishing expedition?
- Not taking into account any major pitfalls, e.g. in relation to the proposed science or technology. It is important to flag up any limitations associated with your proposal and subsequently how you plan to overcome these.
- No credible contingency plans to minimise foreseeable technical issues.
- Science excellence is poorly communicated. You need to write clearly and convincingly.

9.3.3 Impact Is Weak

- Project impact (e.g. IP generation, product commercialisation, uptake by regulators) is not clear, limited or not communicated at all. Remember that impact will be governed by the agency, its mission and requirements, as set out by the call text.
- No clarity on how results are to be exploited beyond the lifetime of the project.
- Project is short-sighted, e.g. no clear link to other current research programmes or initiatives or how the outputs will be of use to other stakeholders.
- Project has little added value to other stakeholders, e.g. society.

9.3.4 Lack of Credibility that the Project will be a Success (Peach, 2017)

- Proposal is not well written, e.g. unclear or confusing sentences, faulty grammar and illogical sentence construction. Remember to avoid opaque language, long sentences, poor headings/subheadings titles, use of lengthy sentences/acronyms/technical jargons, etc.
- Poor organisation of the writing material.
- Bid proposal does not give an impression of professionalism.
- A clear need for rewriting, with a shift in emphasis.
- Proposer does not have anything interesting to say.
- Bid proposal lacks focus, such as the abstract not coinciding with the entire proposal.

- Credibility of the researchers is questionable. For example, relatively unknown institutes or researchers in the early part of their career will often face these hurdles.
- No clear information on project outputs, e.g. milestones/deliverables.
- No clear information on how project outputs are to be disseminated/shared, either within the consortium and to external parties.

9.3.5 Budget Information Not Logical (Lewis, 2007)

- Lack of budget justification.
- Promising to deliver too much or too little.
- Price is not realistic, i.e. a mismatch of price quoted with services offered. For example, there is no explanation as to why the price is too low or too high. You need to state a reasonable budget that is balanced with the work to be delivered within the timeframe proposed.

9.3.6 Management Structure Is Not Communicated Clearly

- No clear and direct line of accountability.
- No process in place in case of potential issues that can slow down the project, e.g. disputes or amongst partners.
- No credible contingency plans to minimise foreseeable management issues.

9.4 Things to Include

Finally, in addition to the above list on what NOT to do, you will need to understand what a reviewer is likely to be looking for. He/she will often be looking for that one proposal that differentiates itself amongst others, thus establishing a clear competitive advantage. There are several ways that you can achieve this (Blackburn, 2003):

- Come up with a great and unique idea.
- Present some (preliminary) promising laboratory results, to give conviction and credibility of the science or technology proposed. This can be presented under the section of *Prior Work* in your main body of text.
- Communicate added value well. Remember that it will be the agency who determines what the added value should be. For example, if the agency's mission is to enhance economic impact, then added value can relate to commercial exploitation, e.g. IP generation and the formation of a spin-out company. On a technical level, added value may involve the adoption of a new scientific strategy or infra-

structure on how to solve an old and continuing problem. One idea may involve the development of a new computational model, to reduce lengthy regulatory testing that will ultimately save time and money in the future. Be specific when communicating added value. For example, if you foresee the generation of a spin-out company, you will need to describe what the commercial product will be. What exactly will the product look like? How much are you planning to sell them for? How will you get the funding to start manufacturing and marketing the products?

9.5 Summary

I wrote this chapter to give you guidance, on how you can improve your success rate to get science funding, which is necessary to keep your research afloat. Although there are different types of bids, this chapter has particularly focused on scientific research grant proposals. The first part of the chapter details on what goes on behind the bid proposal process, in which three main phases have been identified. The second part of the chapter gives practical guidelines on what to write, specifically in relation to some of the common elements often found in a scientific grant proposal. Finally, I have presented some of the common reasons as to why proposals can get downgraded. I have listed these in the hope that you will avoid making the same mistakes, thus improving your chances of success. Finally, remember that bid writing is an invaluable skill, one in which I hope you will develop through time and nurture as your career progresses.

References

Blackburn, T. R. (2003). *Getting science grants: Effective strategies for funding success*. Hoboken: John Wiley & Sons.

Cleden, D. (2011). *Bid writing for project managers*. Abingdon: Routledge.

Foster-Gilbert, C. (2001). *The ethics of medical research on humans*. Cambridge: Cambridge University Press.

Frederick, P. (2011). *Persuasive writing: How to harness the power of words*. London: Pearson Business.

Hubrecht, R. (2014). *The welfare of animals used in research: Practice and ethics*. Hoboken: Wiley-Blackwell.

Lewis, H. (2007). *Bids, tenders and proposals: Winning business through best practice*. London: Kogan Page.

Nickson, D., & BCS, T. C. I. for I. (2012). *Bids, proposals and tenders: Succeeding with effective writing*. London: British Computer Society.

Peach, K. J. (2017). *Managing science: Developing your research, leadership and management skills*. Oxford: Oxford University Press.

Chapter 10
How to Write

For most people, writing does not come naturally. It is rare to find an individual (without adequate training) who can write and produce a polished piece of work in one sitting. Writing can be a huge struggle, as it involves the need to transfer thoughts to paper and convey your message clearly and accurately. When compared to speaking, writing is so much more complex, as it requires a more formal organisation of the text material.

My goal for this chapter is to give you guidance on how to become a better writer. In order to do this, you will first need to understand the writing process, which will be discussed in greater details in the first part of this chapter. In the second part, I will give you practical tips on how to come up with that first draft and how to overcome potential challenges/issues during the writing stage, such as how to overcome a writer's block.

Although the process of editing can be thought of as an activity that is part of writing, for the sake of simplicity, I will differentiate the two and deal with them separately. As such, after reading this chapter, you should continue to read Chap. 11 (How to Edit), to learn how you can transform your first draft into a final version.

On a final note, please do not expect miracles. For sure, you will not become a good writer, just by reading these two chapters. Writing is a skill that you will need to develop further, through time and practice.

10.1 The Process of Writing

As depicted in Fig. 10.1, the process of writing can be divided into three different phases:

R. Tantra, *A Survival Guide for Research Scientists*,
https://doi.org/10.1007/978-3-030-05435-9_10

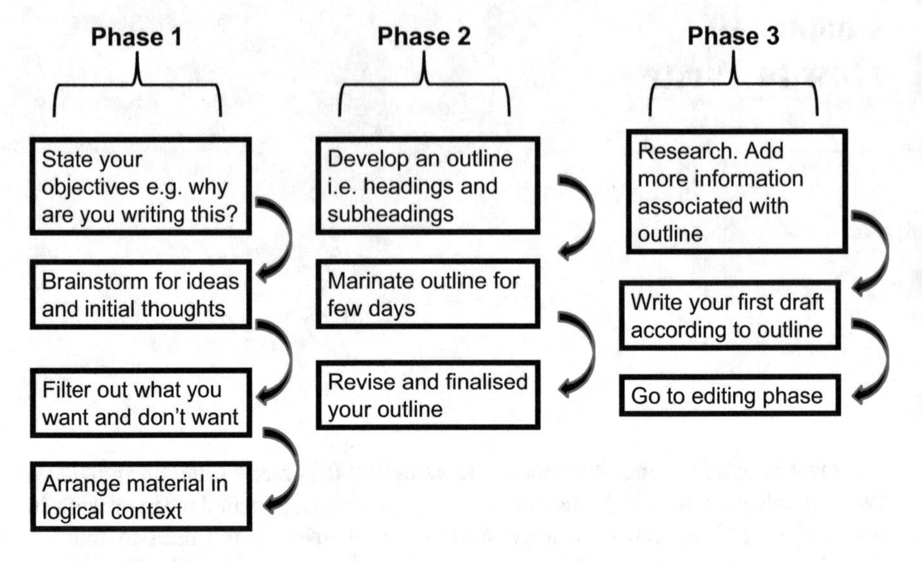

Fig. 10.1 The three phases of writing

1. Phase 1: In this phase, you will need to be clear on your objectives (Forsyth, 2013):

   - Why are you writing this?
   - What do you want to achieve?
   - Who are your readers/audience? What are their backgrounds?

   By answering these questions, you will be able to:

   - Identify/understand your audience and what they expect from you.
   - Establish scope, i.e. what topics/sub-topics to include and exclude.
   - Set the tone and style of writing. Do you need to follow a degree of formality with the writing, a set of rules, style or convention?
   - Establish length of writing. This may be the expected length from your readers or the expected length in which you can successfully convey your message across.

   Your next task is to identify potential headings/subheadings for your piece of writing. In order to do this, you will need to brainstorm for ideas (Clark, 1989; Cory & Slater, 2003). The key to brainstorming is to write down anything that comes into your mind without initial judgement. Once you have a list of ideas in bullet points, you will need to filter, by keeping the ideas that you want and throwing away the bad/weak ones. You will then need to arrange the bullet points in some kind of logical context.
2. Phase 2: The aim of this phase is to filter the outputs from phase 1 and to develop an outline that will serve as a kind of a map for you, when you write. This map is necessary to ensure that your writing stays focused and does not stray into unwanted territories. The kind of outline structure that you will adopt will be

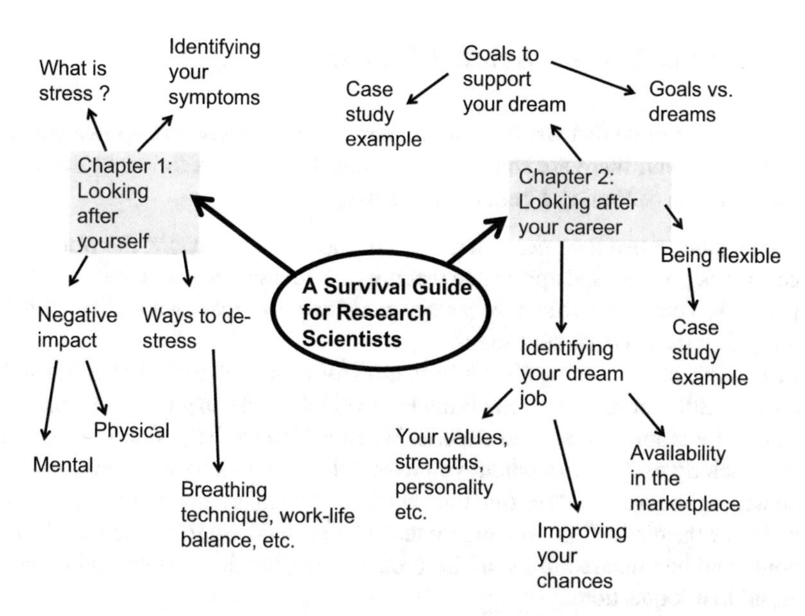

Fig. 10.2 Illustration of (part of a) spider diagram outline to depict a helicopter view of this book

entirely up to you, e.g. spider diagrams, tree diagrams and slides on PowerPoint. Choose an outline structure that works for you. For me, a spider diagram is a great way to allow you to visualise how the different headings/subheadings interlink with one another within a common theme. Figure 10.2 illustrates (part of) a spider diagram that I did before I wrote this book, depicting a helicopter view of the book's contents.

Once you have developed a suitable outline, you will need to leave it alone for a few days, or what I refer to as the *marinating period*. Do not skip this step, as it is imperative that you have a pair of fresh eyes before you edit your outline. It may be that as a result of the marinating period, you will come up with extra headings/subheadings for inclusion. You may also want to rearrange your headings/sub-headings better to improve flow of writing.

3. Phase 3 is all about doing your research, collating the background information that you will need, in order to come up with appropriate text for your first draft (Kaye, 2008). In this phase, you will need to make notes from literature references. Remember that it is quite easy to copy material from literature, word for word. If you are doing this, then you will need to jot down the source details, so that you know where you have copied the material from. When it comes to writing your first draft, remember to read your notes carefully and filter out the information to subsequently interpret your findings. Unless you are actually quoting from a source material, everything that you write must be in your own words. Once you have completed your first draft, you will then be ready for editing (please refer to Chap. 11 on How to Edit).

10.2 How to Become a Better Writer

I would like to iterate that writing is a skill and in order for you to improve you must practice. However, there are some practical tips that you can follow to make your journey towards becoming a better writer easier:

- Do not panic when it comes to writing. The more that you panic, the worse it will get. Think positive and appreciate that it will take time for you to become a good writer. You can also avoid unnecessary panicking, by creating a realistic timeline for your writing (from start to finish).
- Be creative on how you collect information for your background research and be open to different sources of material. For example, you can gain invaluable information by talking directly to experts. When asking for help, don't be scared in approaching people with whom you have not met. Obviously, not everyone will be willing to give you their free time but from my experience some have actually welcome the distraction. Remember that when you do ask for someone's help, be polite and be prepared to listen. Be proactive in your discussions and do not be afraid to ask questions.
- Remember to clarify early on the purpose of your writing and who your target audience will be. This will allow you to identify the kind of language and the level of formality you will need to adopt when you start writing (Van Emden, 2001).
- Read in your spare time and be critical of other people's work (Provost, 1985). Sooner or later you will be able to differentiate between what is good writing and bad writing. Through reading other people's work, you will be able to:
  - Appreciate the different styles of writing.
  - Appreciate why certain style of writing works and why others do not. For example, you may be able to identify how to make your writing more interesting to others.
  - Improve/expand your vocabulary, grammar, etc.
- Practice makes perfect, so write on a regular basis and if possible ask someone else to criticise your work. You will be able to learn from your mistakes and identify your strength and weaknesses as a writer.
- You will need to find a writing routine that works for you, e.g. if you need to write in complete solitude. Through time, you will begin to develop your own routine on how to best tackle the writing process. Your main goal when writing is to find your inner voice!
- Find out if there are other skills that you can learn, in order to support your writing. For example, when I first began writing, I often would use a fountain pen and paper, to write my first draft. This was in the days when I did not have access to a computer and Microsoft Office. However, as computers and software became more accessible, I found it easier to transfer my thoughts straight into the computer rather than writing them out by hand. As a result, I learnt how to touch type. If you are like me, in that you find it difficult to read your own handwriting, then the ability for your touch type will become a necessity.

- Identify any issues that can prevent you from writing. This may be things like physical distractions or background noise. Come up with creative ways to avoid or overcome such problems. For example, if you are in a busy office environment and the background noise disturbs your concentration, then you can use noise cancelling headphones or earplugs (Fry, 2012).

10.3 Having a Writer's Block?

When you get a writer's block, you will find it hard to translate thoughts into words. At worse, you can end up staring into the computer (on a blank page) for hours, hoping for a miracle to happen.

When you experience this, then my first advice to you is to RELAX.

Breath in, breath out, breaaathh away … relax … relax …

Put on some relaxing background music if it helps. In the past, I have used music to help me enjoy the writing process. You can go on YouTube and listen to a classical study music list (or something similar), to help with your concentration. The trick is to find the right type of music for you. Sooner or later, you will forget about the idea that you are actually writing.

When coming up with your first draft, do not worry about the shape of what the writing looks like. Do not worry about the grammar or spelling. Do not worry about the accuracy of the information. JUST WRITE. Write as if you are explaining the material to a colleague in an e-mail—you can be as chatty as you want (Toft & Ellis, 2009). If you find it easier to transfer your thoughts into a series of bullet points to start off with, then do so. Find what is comfortable for you so that you can get those ideas on paper. Be creative on how to solve any dilemmas, to ensure that you do not procrastinate. For example, when I find it hard to express my ideas on paper, I would use a voice recorder to record all of the information. I would then listen to the recording whilst at the same time translating the appropriate text on paper. Overall, remember that the writing experience will be unique to you and you will need to develop a chosen routine that is right for you.

10.4 Summary

Writing is not easy and the goal of this chapter is to arm you with the necessary knowledge so that you can eventually be on the right path to become a great writer. I have given you background information and discussed what goes on behind the writing process (in which three main phases have been identified).

When it comes to the actual writing, I have given you some tips, such as the need to:

- Clarify your objectives; this in turn will be governed by the needs of your readers.
- Develop a suitable outline, to serve as a map and guide you when you do write.
- Conduct a thorough research (prior to writing).
- Filter/interpret the researched material, when coming up with your first draft (everything in your own words, please!).
- Develop a writing routine that works for you.

References

Clark, C. H. (1989). *Brainstorming: The dynamic new way to create successful ideas*. Classic Business Bookshelf.

Cory, T., & Slater, T. (2003). *Brainstorming: Techniques for new ideas*. Indiana: iUniverse.

Forsyth, P. (2013). *How to write reports & proposals*. London: Kogan Page Limited.

Fry, D. (2012). *Writing your way: Creating a writing process that works for you*. Cincinnati: Writers Digest Books.

Kaye, T. (2008). *Learn how to write books that you will be proud to sell*. Morrisville: Lulu.com.

Provost, G. (1985). *100 ways to improve your writing*. New York: New American Library.

Toft, D., & Ellis, D. B. (2009). *From master student to master employee*. Boston: Houghton Mifflin Co.

Van Emden, J. (2001). *Effective communication for science and technology*. Basingstoke: Palgrave.

Chapter 11
How to Edit

Editing is a process of checking through a piece of writing, whilst making revisions along the way, in order to transform a first draft into a final (polished) piece of work (Samson, 1993). The process of editing is not straightforward and often this cannot be achieved successfully in one sitting. Editing is an iterative process and in reality you will often need to edit several times before you end up with a version that you are happy with.

This chapter follows on from the last chapter (that teaches you how to write, in order to create that first draft). Here, I will teach you how to edit and how to make the necessary changes to your first draft. In order to do this effectively, you will need to appreciate the different elements that make up good writing. As such, I will show you how you can edit for:

- Structure and content
- Style and readability
- Spelling and grammar

I will give you tips on how to develop your own editing strategy, i.e. a routine that works for you, whether you are a novice or proficient writer/editor. Finally, the chapter discusses how you can edit other people's writing. The task of editing someone else's work can be a lot harder than editing your own, as you will not know in advance what the writer wants to say.

11.1 How to Edit Your First Draft

The first step in editing is to allow your first draft to *marinate*. By this, I mean leaving the first draft alone and not look at it, preferably for at least a week. This first step is necessary to temporarily detach yourself from what you have written. Thus, when you are ready to edit your material, you will be able to do this with a fresh pair of eyes. Once the marinating period is over, the hard work of editing begins. In order to do this effectively, you will need to edit for several elements, which will be discussed here.

R. Tantra, *A Survival Guide for Research Scientists*,
https://doi.org/10.1007/978-3-030-05435-9_11

11.1.1 Editing for Structure and Content

So, what is the difference between structure and content?

Quite simply, the structure can be considered as is its layout whereas the content is the text within that structure. In other words, you can think of the structure in a piece of writing as the skeleton, whilst the content represents the flesh on the bones.

When you edit for structure (i.e. skeleton), you need to ensure that:

- Titles for headings/subheadings are accurate.
- Headings/subheadings are ordered, so that these flow logically from one to another, so the writing becomes coherent.
- You do not have redundant sections or what I refer to as *padding*, i.e. text that does not seem to add anything. If you have these, it is best to remove them.

If you have developed your outline properly before writing, then the structure of your first draft should be in good shape and you will need to do little to improve this part.

Once you are satisfied with the structure, you can edit the content. In order to do this, you will need to carefully read the individual paragraphs and ensure that (Van Emden, 2001):

- All of the material is relevant. Any irrelevant material should be deleted. For example, if a journal asks you to separate results and discussion into two separate sections, then you will need to make sure that you do not try to interpret or offer opinions in the results section. Another mistake I often see is when people write an abstract that reads more like an introduction.
- You assess your sentences and paragraphs carefully. Check that all sentences belong to a particular paragraph. You may want to delete entire sentences simply because they are redundant, i.e. they do not add anything special. You may even want to delete an entire paragraph and replace it with something else. Make sure that all sentences and paragraphs are arranged in a logical fashion.
- You have explained things sufficiently, i.e. you have included all of the relevant material. Have you missed anything important?
- You are not duplicating ideas, i.e. repeating the same message but in different sections.

11.1.2 Editing for Style and Readability

Style in a piece of writing is all about how you go about saying things (rather than what you actually say). Ensure that your writing style is suited for the purpose in which it is intended for, as there are different styles that you can adopt. You will need to appreciate the kind of style that you may need to adhere to, e.g. what is expected within your own scientific discipline. For most technical writing, the style is usually formal rather than chatty, thus avoiding the use of creative and flowery language. If you are writing a bid proposal, then the style of writing has to be persuasive, in order to convince the readers to buy into your ideas. Whatever style you decide to adopt, you will need to make sure that you do not bore your reader, for example by adopting the same sentence length throughout.

The term readability is a little bit more difficult to define. Overall, it is all about how well your document reads, to ultimately gain the attention of your readers. It is about giving an overall impression that your writing is logical and sensible. There are several tips that will improve the readability of your writing, namely the need to (Cahn & Cahn, 2013; Forsyth, 2013):

- Write clearly. You can do this by:
  - Getting the story right in your head: If you do not fully understand something, then do not attempt to write it; you will only end up confusing the reader.
  - Avoiding vague generalisation: Try to be more specific by giving details and examples, to illustrate your ideas.
  - Adding bullet points, when appropriate.
  - Keeping things simple, brief and to the point: For most technical writing, you will need to adopt a straightforward manner on how to explain things. Do not waffle. In order to convey complex ideas, use shorter words, sentences and phrases. Long sentences can appear woolly or unclear.
- Write accurately. You can achieve this by:
  - Using the right words, e.g. replacing the word *reproducibility* when you really mean *repeatability*.
  - Ensuring that your sentences have accurately conveyed what you initially want to say.
  - Avoiding repetitions: Don't say things twice, e.g. *11 pm in the evening*.
- Structuring text material in a logical fashion, thus achieving flow in your writing: As such, you will need to use clever signposting throughout, to give a logical sequence and shape to the writing material. Signposting is all about flagging to the reader of what is to come, so that the reader knows where it is that you are heading.

11.1.3 Editing for Grammar and Spelling

When I was at school, I was never taught grammar, not formally at least. It is just one of those things that you are supposed to pick up along the way. Although you may not know the correct terminology associated with grammar, this does not mean that you are useless at the subject. The easiest way to learn grammar is to do a lot of reading. By doing so, you will be able to pick up grammar indirectly/ subconsciously.

Although it is not the intention of this chapter to cover all topics associated with grammar, there are some of common dos and don'ts that you should be familiar with (Corder, 2013). If you would like to learn more about grammar, then there are past references that you can refer to for further information, such as Dixon (2011), Dutwin (2004), and Huddleston and Pullum (2005).

Do's and Don'ts of Grammar

- Do choose the correct voice for your writing. Although an active voice is clearer than a passive one, scientific writing often requires that you use the passive voice for reporting, e.g. *the flask was washed* instead of *I washed the flask*.
- Do choose the right tense for your writing, i.e. past/present/future. For example, a past tense is often used to describe a method.
- Do use punctuations correctly. There are some punctuations that you are already familiar with, such as full stops, commas and question mark. However, there are some punctuations that are either being used too often or incorrectly. A common example is the use of colons and semicolons. Colons are often used to introduce a list, a quotation and an equation (with the equation being on a separate line) or to clarify one half of a first sentence. Semicolons on the other hand are used as an alternative to a full stop. Particularly, semicolons are used to join two separate grammatically correct sentences that are closely linked together, with the aim to emphasise the connections between the two. Remember to also use brackets sparingly, i.e. used for abbreviations and the occasional side comments only.
- Do use bullet points to list something, as this will help the reader to scan the material easily. If you add a numbering/alphabetical system to the list instead, then you will be introducing hierarchy into the text, with the first one on the list being the most important.
- Do not use double negatives, i.e. a sentence that contains two negative elements.
- Do not use jargons (slang).

Finally, remember to do a spelling check of your document. When it comes to spelling, remember to (Corder, 2013):

- Choose the right words. There are some words in the English language that can be confusing and often sound similar, e.g. *affect/effect*. There are also some confusion with scientific words that do not sound the same, e.g. *accuracy* versus *precision*, and *reproducibility* versus *repeatability*. If you are unsure to the meaning of a particular word, then you will need to check for its definition.
- American or British English: e.g. words such as *centre/center,* and *colour/color.* Remember to choose one type and be consistent throughout.
- Singular versus plural forms, e.g. *spectrum/spectra,* and *criterion/criteria*. Know what these are and choose the right one.

11.2 Developing an Editing Strategy

A key part of editing is the ability of you to develop your own strategy, i.e. an editing routine that works for you. For example, when editing, do you want to edit for content first or is it best to consider all of the elements (mentioned above) simultaneously? Quite simply, you will need to be flexible in your approach to see what works, as the chosen routine will be governed by your own personal preference and circumstances, e.g. your level of experience as a writer and editor. If you are a novice at writing, then it may be easier to edit one element at a time. For example, you will want to firstly ensure that you are happy with the structure, e.g. to ensure that there is logic in the structure, before moving on to edit for content (to make sure that all the necessary information has been included and that the information is indeed accurate/up to date). Afterwards, you may want to edit for style, e.g. level of formality associated with the writing before proceeding to check for readability and clarity on how well you have conveyed the main message across. You may then proceed to check grammar and spelling, remembering to use the appropriate scientific notations, e.g. how to correctly present molecular or atomic state (CBE Style Manual Committee & CBE Style Manual Committee, 1994). Obviously, once you have become more proficient in writing/editing, you may want to consider more than one element at a time or you may even want to start editing whilst you are coming up with that first draft.

So, what is my personal preference when it comes to editing a piece of work?

For me, the marinating period, in which I need to leave my first draft alone for a week or so, is essential. I then tend to split the editing process into two parts (as illustrated in Fig. 11.1). The first part of editing solely focuses on the structure and making sure that I am completely happy with this before proceeding any further, as

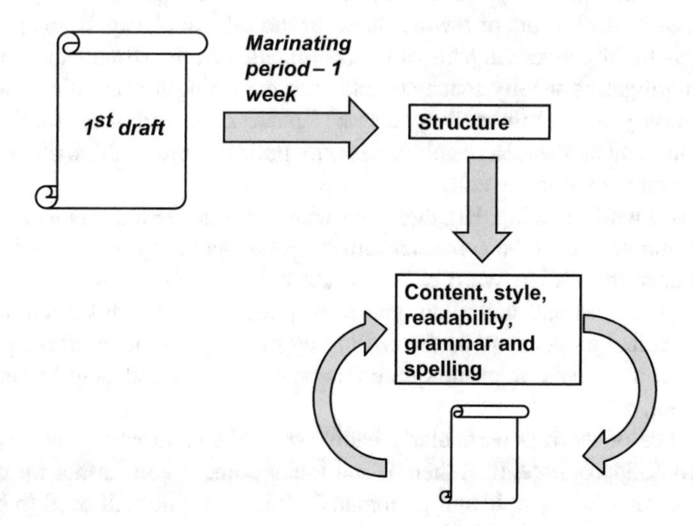

Fig. 11.1 My personal strategy on how to edit a first draft to get to the final, polished version

this is the foundation of the entire story. Once satisfied, I usually proceed to check for all of the other elements (content, style, readability, grammar and spelling) simultaneously. It is on this second part of editing that I tend to spend most of my time, to make the necessary changes. Often, I will repeat this numerous times before coming up with a final version.

11.3 Editing Other People's Work

Let's get one thing straight. You can only edit other people's work if you are familiar with the subject area/topic. If you do not have the sufficient background (in particular the technical expertise), then it would be difficult for you to make comments and edit. Even if you are familiar with the work, editing other people's work can be an uphill struggle, as some scientists can have poor writing skills. Nonetheless, here are some guidelines that you can follow, in order to make your life a lot easier:

1. You need to grasp early on an overview of the entire document. You can develop a sense of this by reading the title and then scanning sections like table of contents, executive summary/abstract, headings/subheadings and conclusion. By doing so, you will have a quick grasp on the key message that the writer wants to convey.
2. Decide early on how much time you are prepared to edit and to what extent. Obviously, if you are a co-author or a supervisor to a student, then you should invest more time and effort, simply because this is expected of you. However, if you are correcting a piece of work for a colleague, it would be a waste of your time to scrupulously edit an entire piece. Instead, it is best to give general feedback on what needs improving and not waste your time going through sentences for grammatical errors or rewrite them for the sake of clarity. If you have time, you can highlight certain parts of the documents that the writer can improve on, e.g. highlighting woolly sentences and recommending for rewriting. No matter how busy you are, minimally you should make some comments on the overall structure and content. Remember, it is only polite to inform the writer as to how much you are willing to edit.
3. Be active whilst reading. Highlight any sentences that are important or those that stand out; do this in the computer rather than by hand, e.g. using track changes in Microsoft® Office Word or Adobe Acrobat DC. Doing this will help you guide your eyes and will allow you to stay focus (to absorb key points). Also make critical judgements on the writing by making side notes whilst you read. You can also make relevant side comments on items that may be unclear or incorrect.
4. If a piece of work is particularly badly written, e.g. structure not logical and content inappropriate, then there is not much point to edit further for elements such as readability, style and grammar. In this case, you will need to hand the writing back to recommend for a complete rewrite.

11.4 Summary

Although editing can be thought of as part of writing, I have purposely dealt with these two topics separately. This chapter, which is solely about editing, thus follows on from the last chapter (that teaches you how to write and how to come up with that first draft). In this chapter, I have identified and discussed the different elements that you will need to consider when you are editing a piece of writing. As such, I have given you practical guidelines on how to edit for structure, content, style, readability, spelling and grammar.

As with writing, my advice here is for you to be flexible in how your approach, so that you can develop an editing routine that works for you. Ultimately, this will be dependent on your personal circumstances, particularly your level of experience as writer/editor. Finally, the chapter gives you tips on what you should do if you need to edit other people's work. Editing other people's work is usually more challenging than editing your own work, as you will not know in advance the overall message that the writer initially wants to convey.

References

Cahn, S. M., & Cahn, V. L. (2013). *Polishing your prose: How to turn first drafts into finished work*. New York, NY: Columbia University Press.

CBE Style Manual Committee & CBE Style Manual Committee. (1994). *Scientific style and format: The CBE manual for authors, editors, and publishers*. New York, NY: Cambridge University Press.

Corder, N. (2013). *A straightforward guide to writing good plain English: Improve your written English*. Brighton: Straightforward Publishing.

Dixon, W. (2011). *Essential elements of English grammar: A guide for learning English*. Verlag: iUniverse.

Dutwin, P. (2004). *Easy English grammar step-by-step*. New York: McGraw Hill Professional.

Forsyth, P. (2013). *How to write reports and proposals*. London: Kogan Page Limited.

Huddleston, R. D., & Pullum, G. K. (2005). *A student's introduction to English grammar*. New York, NY: Cambridge University Press.

Samson, D. C. (1993). *Editing technical writing*. New York: Oxford University Press.

Van Emden, J. (2001). *Effective communication for science and technology*. London: Palgrave.

Summary

References

Part IV
Your Interactions with the Outside World

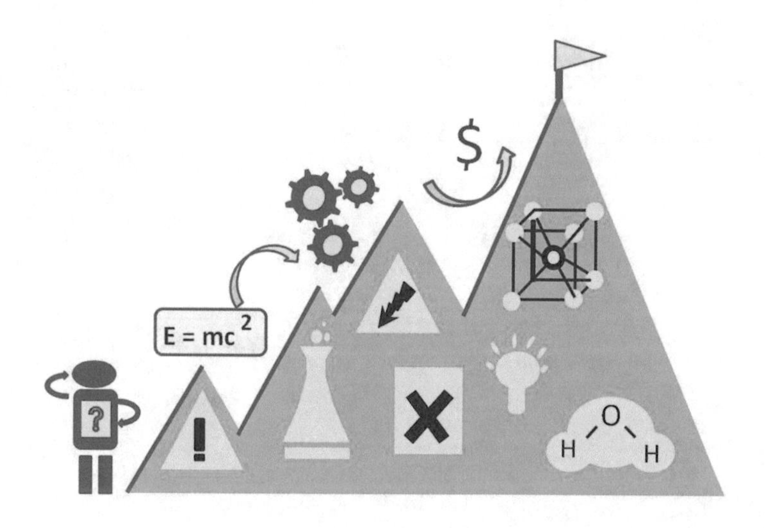

Chapter 12
Meetings

So, how do you define a meeting?

A meeting is a gathering of people with the purpose to hold a discussion for a specific reason; meetings are often based on a set agenda (Raina, 2010). From my own experience of meetings, there were some good ones and there were some bad ones. The bad ones can be a waste of time, leaving you confused and drained. In this respect, I often sympathise when I hear scientists complain that they attend far too many meetings, regretting not being able to invest their valuable time elsewhere, e.g. doing experiments in the laboratory. Although it is likely that you too may have had your fair share of negative experiences, you must remember that meetings can potentially be useful.

In this chapter, I present two aspects related to the topic of meetings: first, what to do if you get a meeting invite and second, what to do if you need to organise one. As such, the first part of the chapter will give you practical guidelines on:

- How to decide whether or not to accept a meeting invite.
- How to politely decline a meeting invite.
- What alternative options you should consider, should you decline an invite.
- How to effectively communicate and behave, when you do attend a meeting.

The second part of this chapter will give you tips on how to organise for a meeting. In particular, I will give you tips as to:

- Who you should invite.
- How you can draw up a suitable meeting agenda.
- How you can write up meeting minutes.

R. Tantra, *A Survival Guide for Research Scientists*,
https://doi.org/10.1007/978-3-030-05435-9_12

12.1 Meetings: To Go or Not to Go?

When you have a received a meeting invitation, you will first need to decide if you want to go or not. In order to establish this, you must first understand the purpose of the meeting. If you are unsure, then you must contact the meeting organiser or chairperson, to seek further clarification.

There are several reasons as to why people will want to have a meeting, to include the need to (Raina, 2010):

- Brainstorm, to find a creative solution to a problem.
- Report on recent activities, e.g. progress on projects and plan for future ones, which will help to clarify duties/accountabilities.
- Assess performance, e.g. staff appraisals.
- Negotiate, e.g. your salary.
- Deliver a message to a wide audience, e.g. informing employees of planned redundancy.
- Etc.

Once you have understood the purpose of the meeting, you will need to decide if you are going to attend. You will need to make a list, to weigh out the pros and cons (as illustrated in Fig. 12.1). Remember that the benefits of attending a meeting may not be so apparent at first. For example, there are several added values associated with a meeting that you may not be initially aware of, such as the need to:

- Promote staff bonding/team spirit
- Sustain and expand your network/contacts
- Gain valuable information, e.g. from other experts

If you see little benefits in attending a particular meeting, then you must contact the organiser or chairperson to flag up your concerns. Explain to the organiser, as to why you may be reluctant to attend. Remember communication is key. If you are declining the invite, then you must provide a good enough reason, such as (Lomas, 2005):

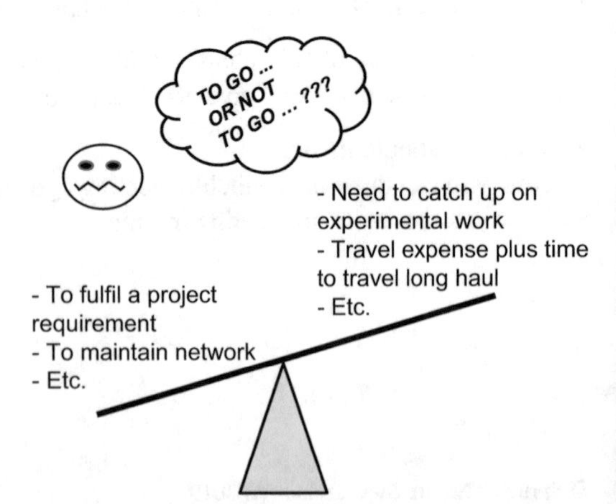

Fig. 12.1 Considering the pros and cons prior to accepting a meeting invite

- The inability to justify the costs of time and travel
- The need to prioritise a heavy workload
- The need to attend another meeting
- Your unavailability workwise, e.g. on annual leave, and surgery

Before you do decide to fully decline a meeting invite, you may want to consider other options. For example, if you want to attend the meeting but cannot justify travel or time costs, then discuss with the meeting organiser the possibility of (Lomas, 2005):

- Joining the meeting through a tel-con and attending part of the meeting and leaving when you are no longer needed
- Being replaced or represented by someone else
- Getting in touch with the organiser after the meeting, e.g. by sending an e-mail to discuss any major outcome and asking the organiser to give you a copy of the meeting minutes
- Etc.

12.2 How to Organise a Meeting

If you have been asked to organise a meeting then it is your responsibility to decide on (Weynton, 2002):

- Location and time.
- Size of room needed.
- Who is going to cover the cost of the meeting.
- Whether or not food/drink will be provided. And if so, who will provide this?
- Contents of meeting schedule or agenda: Remember that when drawing the agenda, you will need to allow enough time for participants to reach the meeting venue.
- What supplies to provide in the meeting room, e.g. computer, pen and papers.
- What technology set-up to provide, so that those who cannot be physically present at the meeting can also take part: The need to have cameras and microphones is necessary if you want to allow attendees to join via a tel-con, e.g. via Skype meeting ("New Skype | Enhanced features for free calls and chat", n.d.) or GotoMeeting ("Easy Online Meetings With HD Video Conferencing | GoToMeeting", n.d.).

As the organiser, you will need to determine who you will want to invite. You will need to identify the key people that should be there. In order to establish this, you must understand the different roles that different key players can play (Burns, 2002):

- Participants: The role of a participant is to actively participate in a meeting and his/her voice will influence the outcome of that meeting, e.g. influence important

decisions made. It is wise to invite different people, in order to have a more balanced debate/outcome. Hence, consider inviting those that may have strong or negative opinions, as they can act as a good sounding board to ensure that the best decision will be made.
- A chairperson: The role of the chairperson is to steer the meeting, making sure that the discussion stays focused. Skills needed for this position include diplomacy, leadership and being a good communicator. Like a judge, the chairperson should adopt a neutral attitude and not contribute heavily into the discussion.
- A minute/note-taker: It is this person's responsibility to take notes during the meeting. He/she is responsible for writing the minutes at the end of the meeting, for distribution. If the meeting is small, then there is no need to have a dedicated note-taker as either the organiser or the chairperson can also fulfil the role quite easily.
- Facilitators: Facilitators are chosen to help out during the meeting. For example, facilitators may be needed in the event of a breakout session, in which specific tasks will be given to different groups. The role of the facilitator in this case is to lead the discussions within his/her own group and then to summarise/present the outcome to others in the meeting.
- Observers: Observers are often needed if the meeting is a formal one. The role of an observer is to oversee the discussions or proceedings that have taken place during a meeting. It is the observer's job to acknowledge that the meeting has been conducted in accordance to the meeting's rules and regulations. Observers will not be involved in the discussion and will not have the right to influence decisions.

As a meeting organiser, you will need to decide who your key people will be. If for some reason some of the key people that you initially identified cannot attend, then consider replacing them with other suitable candidates or else you will need to move the meeting to a different date and/or location.

So, who NOT to invite?

Quite simply, you should not invite those that have little or no contribution to the discussions in the meeting. If you think that someone may be able to contribute to parts of the meeting, then consider the possibility to allow him/her to attend at specific times.

As an organiser, you will need to write the meeting's agenda. An example template for the agenda is shown below. In drafting the agenda, you will need to include essential information, such as (Barker, 2011):

- The meeting title, location, date.
- The purpose of the meeting.
- Name of key participants, e.g. chairperson and facilitator.
- If refreshments will be provided.
- The agenda items (e.g. a welcome speech, introduction of attendees, questions and answers, apologies of absence, breaks, key note speech, oral presentations, panel question/answer session). Ensure that the items of greatest priority are

listed first on the agenda. Remove any unnecessary agenda items. Ensure that there is a timescale associated with each agenda item, i.e. start/end time.
- Clarity on what is expected from the participants, e.g. what they have to do during breakout sessions.
- Any supporting documentations, e.g. map on how to get to the venue.
- Any matters or action points arising from the previous meeting.

Name of meeting
Date of meeting
Location of meeting

Purpose of meeting *"This meeting has been called following ... The meeting is open to any interested party. Participants are requested to fill in the registration form provided. Please refer to attached documents for logistics and other supporting information ..."*
Key attendees (with names/e-mails)
Chairperson: Dr. ...
Secretary: Mr. ...
Contact person: Ms. ...
Meeting agenda
Item 1: (9:50–10:00) Registration ...
Item 2: (10:00–10: 15) Welcome message/opening remarks (chairperson) ...
Item 3: (10:15–10:45) Keynote speech by ...
Item 4: (10.45–12:00) Presentation by ...
Item 5: (12:00–13:00) Lunch
Item 6: (13:00–14:00) Breakout session ...
Item 7: ... Etc.

Once you have written the agenda, you will need to send the draft agenda to the chairperson for approval before finalising and distributing to potential attendees.

12.3 Attending a Meeting

So, let's say that you have been invited to a meeting and you have accepted the invite.

Now what? What do you need to do to prepare? How should you behave during the meeting?

Quite simply, this will all very much depend on the role that you are going to adopt. To illustrate this point further, let us focus on three type of attendees: participants, chairperson and note-taker.

12.3.1 *Participant*

If you come as a participant you will need to:

1. Make the necessary plans (e.g. transport, hotel accommodation) in order to get to the meeting venue.
2. Prepare and read up on any relevant background material beforehand. Remember to read any documents given to you. Plus, do your own research, if necessary. For example, you may want to know more about the participants who will be attending. As such, you can do a background check of key individuals, to understand their interests.
3. Assess the agenda carefully and identify what to expect from the meeting and what you consider to be as a favourable/realistic outcome.
4. Assess your position. For example, if you think that your voice is in the minority (e.g. you will be coming to the meeting with opposing or negative views), then it may be worthwhile to share these views with fellow participants ahead of the meeting, to lobby for support.
5. Ensure that you behave professionally during the meeting. As such, you will need to (Lomas, 2005):

   (a) Stay focus and remind yourself the objectives of the meeting. For example, are you there to help resolve a conflict or solve a problem?
   (b) Identify the message you want to relay across. It may help to write them down in bullet points ahead of the meeting.
   (c) Deliver your message effectively. You can do this by getting people's attention to get your point across and not delivering your key message at the end of the meeting. It is often difficult to reverse or challenge any decision when a meeting is closing.
   (d) Adopt a positive body language. Be assertive, not aggressive.
   (e) Be logical and not emotional, e.g. not to take things personally, especially if your points have been dismissed.
   (f) Be creative and come up with potential solutions.
   (g) Consider the positive and negative implications from any decision, in order to select the best one.
   (h) Listen actively to what others have to say, particularly if they offer opposing views. Do not dismiss opposing views straightaway.
   (i) Ensure that you are clear about any task that are given to you, e.g. during a breakout session.
   (j) Take notes of any plans that are put in place at the end of the meeting, e.g. future actions. Jot down the details: What, when and by who?
   (k) Raise any issue with the chairperson, giving recommendations if appropriate, e.g. how future meetings can be improved.
   (l) Appreciate the fact that conflict can arise. Always be positive and remember that conflicts can be useful, as it adds a different point of view to the discussion. Unfortunately, some conflicts can manifest to personal attacks

and take on an aggressive tone. If this is the case, then you will need to involve the chairperson to help mediate the situation. Having a plan on how you are going to deal with a conflict is always useful.

12.3.2 Chairperson

If you are the chairperson in a meeting, then it is your job to (Clarke, 1993):

1. Follow points 1 to 3 above (as in the participant section).
2. Open and close the meeting.
3. Deliver a welcome speech, to highlight the purpose of the meeting and its importance.
4. Ensure that everyone knows each other, along with their respective expertise. If needed, a *tour de table* should be carried out, especially if the participants have not met before.
5. Encourage active participation from all and ensure that different point of views have been heard. To encourage participation, you will need to create a friendly, yet professional atmosphere. You will need to let everyone know that their input is valued and to ensure that the meeting is not hijacked or monopolised by certain individuals (or groups) and that no one gets bullied.
6. Know what to do should a conflict arise. If it does arise, you may want to first understand the source of the conflict and establish if there are any personal (hidden) agendas associated.
7. Ensure that conversations stay focus. If the discussion starts to divert and become irrelevant, it is your job to remind the participants of the objectives.
8. Know what to do in case of a complaint. First do not dismiss the complaint. Instead, explore how such complaints can be turned into something more positive such as incorporating them into future objectives.
9. Summarise the discussion either at the end of each agenda item and/or at the end of the meeting. When you are closing the meeting, ensure that you give a summary of the meeting outcome and say if the objectives have been met. Also, if relevant, ensure that a casting vote has taken place to identify future actions.
10. Thank the participants for their contribution at the end of the meeting.
11. Ensure that the note-taker has all of the information he/she needs in order to write the minutes.
12. Approve the draft minutes from the note-taker before its circulation.

12.3.3 Note-Taker

If you are a note-taker, then your job is to (Delehant & Von Frank, 2007):

1. Follow points 1 to 3 above (as in the participant section).

2. Know all the people in the attendance sheet, e.g. who they are and which organisation they represent.
3. Record the meeting. Sit next to the chairperson, if possible. Use a computer to take notes and identify key points arising from the meeting. It is useful if you can touch type. If your typing speed is slow, then make sure that you have a pen and paper to take notes. If you have trouble taking notes from a meeting, it might be easier to tape the entire meeting. However, make sure that you seek the permission from the participants before doing this.
4. Actively listen to the discussions; do not listen selectively. Treat everyone with the same degree of respect. If at any stage you are unclear about the points that have been made from the attendees, you will need to interrupt when appropriate and ask for clarification. It is important that you write down the action points from each agenda item, to be implemented in the future.
5. Summarise your notes and draw up the minutes of the meeting (see below for further information).
6. Send the minutes of the meeting to the chairperson for approval, before circulating to everyone else.

12.4 Writing Meeting Minutes

As discussed above, if you are a note-taker, then it is your responsibility to write the minutes up. When you are back at your desk, read through your notes and make sure that you are clear on the different points. If unclear, you can either ask the relevant participant/s or the chairperson for further clarification.

When writing up the minutes, do not waffle or use long sentences. Avoid using jargons and acronyms, in order to make things clear. Consider incorporating the following elements into your meeting minutes (Gutmann, 2013):

1. Title of meeting, date, location.
2. List of attendees, noting down the roles if relevant, e.g. chairperson and observer.
3. List apologies of absence.
4. Confirm any actions from the last meeting and if these have been delivered.
5. State the objectives of the meeting and deduce if these have been achieved.
6. Record what was discussed during the meeting, by whom.
7. Summarise the outcome from each agenda item.
8. Record what decisions have been made, e.g. if consensus has been reached.
9. Identify future action points. You will need to write what these are, who will be doing them and by when. In describing when, use dates rather than something vague like *as soon as possible*.

Minutes should be written in a logical order, i.e. mirroring the order of the items in the agenda (see below for an example of how minutes can be written). If the meeting had been particularly complex and lengthy, then you might find it useful to create an outline first (e.g. using a spider diagram) before writing your first draft. It is best to write the minutes of the meeting straight away while everything is still fresh in your mind.

Meeting Minutes

Name of meeting …
Date of meeting …
Location of meeting …

Purpose of meeting This meeting has been called following …
Key attendees (with names/e-mails)
Chairperson: Mrs. …
Secretary: Mr. …
Contact person: Ms. …
Minutes items
General remarks:

The meeting started at … with a welcome message and opening remarks from the chairperson … There were 40 attendees, in which they all have signed the registration form as proof …

Item 1: Keynote speech was given by Dr. … He highlighted the main issues … He noted that the following considerations should be taken into account during discussions to follow …

Item 2: The chairperson initiated an open discussion after item 1 on the agenda … The outcome of the discussion was …

Item 3: The proposed resolutions resulting from Item 2 are … These have been agreed upon, pending some minor adjustments made to the text. A couple of minor issues came up; these are … As a result, some amendments have been made to …

Etc. …
Action points

Document A should be revised accordingly by … with an agreed delivery date of …

Next meeting to be held at …
Etc.

12.5 Summary

Nowadays, it is becoming more common for research scientists to be invited to lots of meetings. Although meetings can be useful, there are a number of different reasons as to why you might want to attend. In this chapter, I have discussed the pros and cons of going to a meeting and what you should do if you decide to decline an invite. Furthermore, I have presented practical guidelines, listing some general dos and don'ts on what to do and how to behave in a meeting environment. Finally, this chapter gives you guidelines on what you should do if you ever need to organise a meeting. The discussion has particularly focused on who you should invite, how to draw up a meeting agenda and how to write up meeting minutes.

References

Barker, A. (2011). *How to manage meetings*. London: Kogan Page.

Burns, R. B. (2002). *Making meetings happen: A simple and effective guide to planning, conducting and participating in a successful meeting*. Crows Nest: Allen & Unwin.

Clarke, J. (1993). *Managing together: A guide to working effectively as a committee*. Dublin: Combat Poverty Agency.

Delehant, A. M., & Von Frank, V. (2007). *Making meetings work: How to get started, get going, and get it done*. Thousand Oaks, CA: Corwin Press.

Easy Online Meetings With HD Video Conferencing | GoToMeeting. (n.d.). Retrieved November 29, 2017, from https://www.gotomeeting.com

Gutmann, J. (2013). *Taking minutes of meetings*. London: Kogan Page Limited.

Lomas, B. (2005). *Easy step by step guide to fewer, shorter, better meetings*. Utgiven: Summersdale Publishers Ltd.

New Skype | Enhanced features for free calls and chat. (n.d.). Retrieved November 29, 2017, from https://www.skype.com

Raina, R. L. (2010). *Communication for management*. Kolkata: World Press Private Ltd.

Weynton, B. (2002). *Organise meetings BSBADM405A*. Chatswood: Software Publications.

Chapter 13
Presentations

Throughout your career as a research scientist, it is inevitable that you will be asked to deliver a number of presentations. For example, if you attend a conference, you may be asked to present a lecture or a poster. Given the choice, most (nervous) students will opt to do a poster instead of an oral presentation, as the former is a lot easier to deliver. This is ashamed, as giving an oral presentation will give you also an opportunity to excel and be noticed by your scientific community.

In addition to giving an oral presentation in a conference setting, you may be asked to present for other reasons. For example, you may need to (Abrams, Clapperton, & Vallone, 2008):

- Report progress in your company, e.g. to a boss or group/team.
- Disseminate your experimental findings to other scientists in a project meeting.
- Take part in seminars/workshops, communicating with groups of people who share similar interest or those belonging to a specific scientific community. If you are an expert in your field, then you may be asked to present something for the purpose of education or skill development. You may also then be invited to join a panel discussion to share your views (or to answer questions), along with other experts.
- Do a sales pitch, in order to help your organisation to sell products/services to customers.
- Find investors to finance your ideas or set up a spin-out company.
- Etc.

Whatever your reason, it is vital that you have the necessary skills to know how you can deliver your presentation effectively.

The goal of this chapter is to give you knowledge on how to prepare for an oral presentation and how best to deliver it. I will aim to answer the following questions:

1. What is a presentation?
2. What are the phases involved in the preparation and delivery of a presentation?

R. Tantra, *A Survival Guide for Research Scientists*,
https://doi.org/10.1007/978-3-030-05435-9_13

3. How do you engage with the audience?
4. How do you handle nerves?
5. How do you handle question-and-answer sessions?
6. How can you improve as a presenter?

13.1 What Is a Presentation?

In the broadest sense of the word, a presentation is all about human-to-human communication (Bradbury, 2010), with the aim of delivering a message to an audience. Unlike making a speech in social events (e.g. wedding), scientific presentations (in either a formal or an informal setting) will require you to relay accurate, technical information.

As a presenter, you will need to know how to deliver your speech in the most effective way. For you to do this, it is important that you understand and appreciate the four phases associated with doing an oral presentation, namely preparation, planning, construction and delivery (as depicted in Fig. 13.1).

13.2 Phase 1: The Pre-preparation Phase

The aim of the pre-preparation phase is to do research and gather information and do thorough research. As such, you will need to (Peel, 1998):

1. Be clear on the logistics, e.g. where you have to go to present and how you plan to get there.
2. Be clear on the topic that you are asked to present. Quite often, the topic is determined by the co-ordinator or meeting organiser who has invited you.

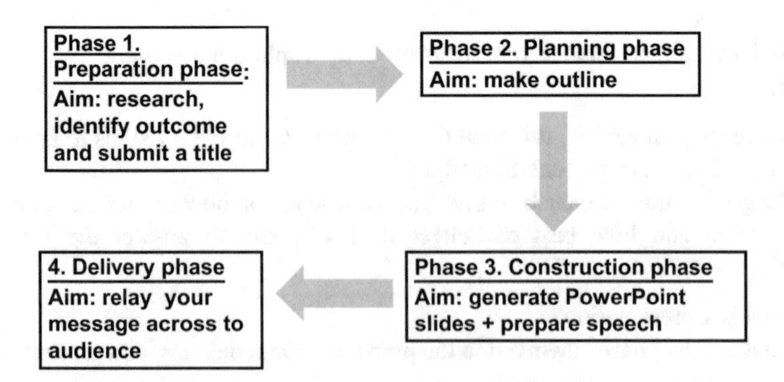

Fig. 13.1 The four phases behind the planning and ultimate delivery of a presentation

3. Understand your audience: By this I mean knowing their profile, which will allow you to define scope for your presentation. Who are they? What are their interests? What do they want to hear? For example, if your audience are experts in your chosen discipline, then there is not much point in presenting a lot of background information. Instead, your focus should be on new, hard data and the specific details associated with your experiments.
4. Identify your objective/s well in advance. Be clear on:

   (a) The core message that you want to convey and how much time you have to deliver.
   (b) The ideal outcome: What do you ask from your audience? What do you want them to do/feel after hearing your talk? Do you want to influence the outcome for a casting vote? Do you want your audience to support your research idea?
   (c) The added value associated with giving your presentation. For example, if you are giving a presentation to report on the general progress of a project that you are leading, then this may be the chance for you to show off your knowledge, e.g. in front of the boss. You may want to portray yourself as being a competent member of staff, who is worthy of a promotion.

5. Be aware of potential troublemakers in the audience and try to identify who they are beforehand. I do not want you to be paranoid but you have to remember that humans are not created equal and as discussed in Chap. 15 (Dealing with Difficult People) human behaviours can be complex and sometimes unpredictable. My advice to you is to just be aware of potential troubles, e.g. office politics. By being aware and anticipating potential issues in advance, you will be in a position in which you are able to accept the situation early on and not take unfair criticisms personally (if it does happen).

At the end of phase 1, you will be in a position to submit the title of your presentation to the organiser. Choose your title carefully, not too long or too short. It should be informative and accurate. If you are going to deliver a presentation for a conference, then quite often you will be expected to submit not only the title but an abstract as well.

13.3 Phase 2: Planning Phase

In this phase, you will need to gather your material and decide on the content of your presentation. Based on the objectives (as identified in phase 1), the aim of phase 2 is to identify what key points/sub-points you will want to include in the presentation that will support your the core message. If your talk is particularly complex, you are likely to have lots of sub-topics interlinked with one another. As such it is best to represent your outline through the use of a spider diagram (as exemplified in Fig. 13.2). Remember not to have too many main points in your

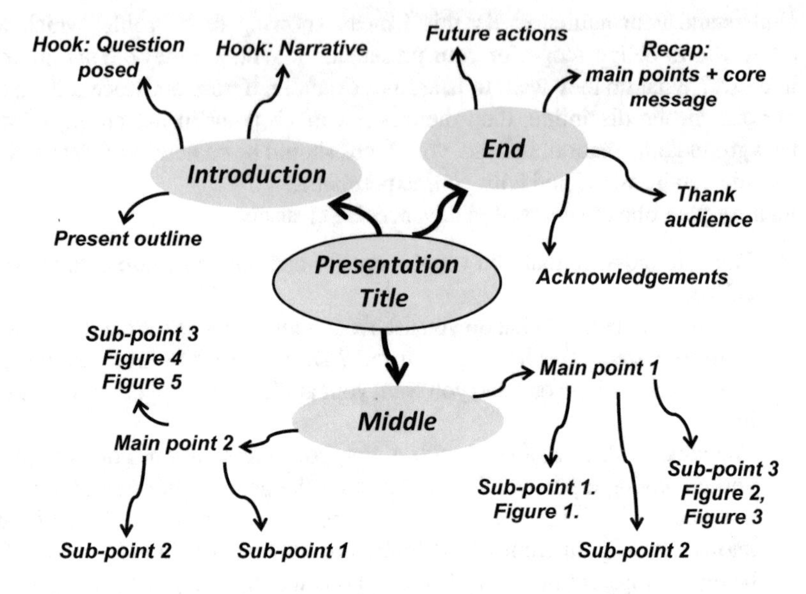

Fig. 13.2 A spider diagram template outlining the points and sub-points to support your core message

outline. If you choose to have too many main points, then the audience will end up with information overload and get bored easily. Ideally, you will want to have two to three main points to support your core message (for 30-min presentation).

In your outline, you should include the following elements (Harvey, 2008; Walker, 2015):

1. *An introduction* section: The introduction has to be seen from the audience's perspective (rather than your own). In other words, your introduction should connect with your core message and what the audience want to hear. By doing so, you will be able to hook your audience, so that they will want to listen to you further. There are various ways in which you can do this:

   (a) Ask a question.
   (b) Make a controversial statement.
   (c) Tell an interesting personal story.
   (d) Do a demonstration, e.g. showing what a particular equipment/technology can do.

2. *A middle* part: This is when you present a collected body of evidence that will support your core message and help to influence the ideal outcome. The body of evidence will need to consist of hard facts, such as technical data, experimental observations, published findings, testimonials and case studies. In presenting your slides, you will need to make a smooth transition from one topic/sub-topic to next, so that there is a logical transition or flow of ideas from one slide to the next, until you reach the end.

3. *A conclusion*: At the end of your presentation, you will need to recap on the main points to support your core message, to ultimately affect the outcome. When drawing up the conclusion, ask yourself:
   What points do I want the audience to remember, such that it will lead to the kind of outcome that I want?
4. *An acknowledgement* section, if relevant.

13.4 Phase 3: Construction Phase

Once you have finished with a suitable outline (as detailed in phase 2), you will then need to construct visual aids to help deliver your presentation. There are different types of visual aids that you can adopt, e.g. flipcharts, white board and videos. However, the standard for most scientists nowadays is to employ PowerPoint slides. If you do not know how to do a PowerPoint presentation, then I strongly recommend that you learn (Lowe, 2016). It is easy to use and fast and will allow you to display your data/information in a professional manner.

Although I will not cover the topic of PowerPoint in great details here, it is worthwhile to consider the following tips when designing your PowerPoint presentation (Jay & Jay, 1999):

- Keep the overall style and design as simple as possible, in order to display your data and information. Do not be tempted to play around with colourings or special effects, e.g. Word Art. Using over the top graphics effects can look silly with a technical based presentation and will distract the audience.
- Avoid too much clutter, i.e. cramming too much information on any given slide. If your slide is cluttered with too much information, then the audience will find it difficult to concentrate on the main point/s. As a result, your efforts in order to get your message across will be severely dampened.
- When creating your slides, always refer to your outline structure (e.g. spider diagram) for guidance. This will ensure that you remain focused on the core message.
- Ensure that you have access to the relevant material, so that you can import data and information easily when constructing the slides.
- Ensure that all text on the slides is readable, e.g. text font not being too small.
- Ensure that you do have meaningful titles/sub-titles on top of each slide.
- Insert any relevant data, graphics or images where needed.
- Ensure that ideas flow logically from one slide to the next. Remember, in effect, you are telling a story from the introduction to the conclusion.
- Make sure that you are confident with the material presented on every slide and that you can comfortably elaborate on them. As such, you will need to take care when recycling old material or other people's material. Make sure that you update slides as appropriate, to better integrate them into your presentation and to support the core message.

- Edit your first draft. Delete any unnecessary items, e.g. images or redundant slides. Ask yourself: Are the slides sufficiently convincing to support the core message and influence the outcome? Will the slides be able to deliver the core message in a simple and clear manner?
- When you have come up with the final version, remember to make a copy of your presentation in a memory stick, to take along with you to the meeting. You can also e-mail the presentation to yourself for extra backup.

13.5 Phase 4: Presentation Delivery

This phase is all about coming up with a speech in order to guide your audience through the PowerPoint slides, to subsequently deliver your core message effectively. You will need to come up with the appropriate text in order to explain the points and sub-points highlighted on the slides. Make sure that you are able to explain things in a clear, simple manner and always use everyday language.

With regard to whether or not you should have notes whilst presenting, this will be a matter of personal choice. If you do decide to have notes, then my advice is to write them on a piece of paper or a few index cards. You can write down key words (in bullet points) to help jog your memory on what you want to say. You can opt for a spider diagram for a more pictorial representation, if this will help. Whatever you decide to do, remember NOT to write the entire text of your speech, bring it to your meeting and read it out loud, word per word. You NEVER want to bring a prepared script. Your aim is to deliver the speech as naturally as possible, as though you are having a one-to-one conversation with someone.

Personally, I do not take any notes with me as I do not like to hold things in my hands and be distracted by having to read them. My personal strategy to deliver a speech is as follows:

1. Have printouts of the slides in advance and take them to the meeting venue. I usually have four slides printed per A4 side.
2. An hour before presenting, you will need to look at the printouts and remind yourself of the content and on what you want to say. I do this by scribbling some notes on the actual printouts. For example, next to the title, I would write a piece of text to remind me of the core message that I want to relay. Then, next to the other slides, I would write in bullet points the main point/sub-points that I want to convey to the audience. The whole process usually does not take me more than 5 min to complete. After this exercise, I will have all the information fresh in my head. This will allow me to deliver my speech with confidence, without the need to resort to any notes.

Once you have worked out on what you want to say, you will need to work on how you are going to say it, as this is key to an effective presentation. There are several tips that you can consider in order to achieve an effective delivery (Chalmers, 2016; Dunn & Goodnight, 2016):

- Be direct, clear and persuasive. Do not waffle.
- Make sure that you master the material, particularly the science that you will be communicating. You can rehearse in front of an audience beforehand. You will want to choose an audience who will be highly critical of your work, so that you can ultimately identify and address any weaknesses/deficiencies in your presentation. Such a practice run will allow you to further edit your slides, improve content quality, improve on speech delivery and ensure that you are indeed sticking to the time limit given.
- Present your speech in a natural manner and use everyday language. Talk as if you are having a one-to-one conversation, rather than addressing a huge audience. Remember that you can only do this if you relax. Please refer to the information below for some helpful hints on how to relax before your presentation.
- Be aware of your body language. Find your unique style to deliver with confidence and a sense of authority, being self-aware on how your voice projects in relation to volume and pitch. Be aware of your speed, as it is likely that you will talk too fast if you are nervous. Remember to slow down, so that you can articulate the words clearly. Introduce strategic pauses, in order to make an important point.
- Engage with the audience, by keeping them interested in what you have to say, as much as possible. You can do this by:
  - Maintaining eye contact with the audience: Remember to shift your gaze at different sections of the audience from time to time.
  - Adopting a relaxed body language to deliver your speech: The more relaxed you are, the more relaxed and receptive your audience will be.
- Ensure that you come in a suitable attire, by dressing appropriately for the kind of meeting that you attending.
- Remember to thank your audience for listening and welcome any feedback.

13.6 Handling Nerves

Although it is quite natural for you to be a little bit nervous, nerves can actually get the better of you (if not controlled properly). To keep your nerves under control, consider the following tips:

1. Accept the fact that you are likely to be nervous, especially if you have not had much experience with delivering a speech in front of an audience. Take comfort knowing that the more you practice, the easier it will get.
2. Understand that when you are nervous, you are likely to be experiencing symptoms of anxiety. You will first need to identify the associated symptoms before you can implement a strategy to control it. Does your hands get cold? Do they shake uncontrollably? Does your voice quaver when you are nervous? Then ask yourself if there is anything that you can do to alleviate the symptoms. For exam-

ple, you may like to adopt a breathing exercise to calm your nerves or visualising something relaxing, e.g. blue sky or a beach with palm trees (see Chap. 1 for more tips on how to relax). If you suffer from cold hands (like me) as a result of anxiety, then it might be worthwhile to take something along with you to the meeting venue, which will allow you to wrap them up in something warm. I would always have jacket with side pockets in it, to warm up my hands (in case these get cold). My point is to be creative in finding solutions in order to reduce feelings of anxiety.
3. Ensure that you practice what you are going to say a couple of times before the actual day. As mentioned above, the need to do this in front of a critical audience at least once is advisable.

13.7 After Your Presentation

13.7.1 Questions and Answers

At the end of your presentation, you should welcome the chance to take on any questions that the audience may have for you. This is your chance to clarify your core message, to explain any points/sub-points that were unclear or needed further verification. In addition, it is always helpful to get feedback. It may give you new ideas on how to proceed with your research.

It is better to handle any questions at the end of your presentation (if possible), to avoid unnecessary distractions during your talk. If you choose to answer questions as you go along, then ensure that you do not get side track by answering questions that have little or no relevance to your core message.

When you are faced with a question-and-answer session ensure that you (Davis, Davis, & Dunagan, 2012):

1. Start with a positive attitude. Remember that questions will only help you to improve, e.g. to further mould your research (career).
2. Listen carefully to what is being asked. If in doubt, ask for the question to be repeated or ask for further clarification. It is imperative that you understand the question before you start formulating the answers.
3. Pause before you answer a question. Do not answer straight away. This will allow you time to think things through and structure your answers carefully before you open your mouth.
4. Come prepared, should you find it difficult to answer a particular question. First, stay calm and do not try to bluff your way out. If you have a colleague who is an expert, then you can divert the question to him/her. If you are really struggling, then admit your ignorance. However, end it with a positive note by asking the person for his/her e-mail address so that you can let the person know of your answer in due course. Always strive to be helpful.

5. Maintain a professional outlook throughout, especially if you have to handle bad behaviour (e.g. which may be due to office politics). Avoid making sarcastic comments, looking annoyed or taking on an aggressive/defensive tone. Try to rephrase the question, so as to reduce any mounting tension. Put everything into perspective and answer the questions as best as you can. Please refer to Chap. 15 (Dealing with Difficult People) for further information.

13.7.2 Feedback

Remember that you will only get better if you can learn from your mistakes. The need for improvement should be a continuous process (throughout your career). Just remember that the more you practice, the better you will get. Hence, at the end of each talk, it is beneficial to gain feedback, to assess how well you did and how you can improve your future performance (as depicted in Fig. 13.3).

If possible, you can ask a colleague in the audience to make notes and judge your performance against various elements, such as (Ledden, 2013):

- The content of your presentation: Have you structured the presentation in a logical manner, so that it can be followed easily by a listener? Did your ideas and sub-ideas flow logically throughout? How well have you prepared the PowerPoint slides?
- Your communication style and how well you have relayed your message across. Were you audible and did you convey your message clearly? Were you passionate in your delivery or did you sound boring? Did you waffle? Did you talk too fast such that it is difficult to understand what you were trying to say? Did you manage to deliver your presentation naturally and convincingly?

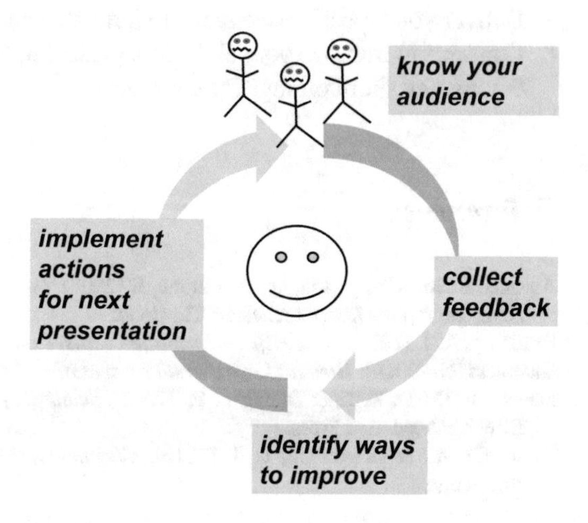

Fig. 13.3 Welcoming audience feedback to improve your craft as a presenter

- How well have you answered the questions at the end of the presentation? Did you welcome people's comments and feedback? Or did you seem uncomfortable and defensive? Did you fully listen to understand what was asked of you? Or do you have a habit of answering the question before a person has finished speaking?
- How well have you dealt with bad behaviour (if any)? When you have been unfairly criticised by certain members of the audience, did you handle the situation with calmness and grace? Or were you sufficiently provoked such that you answer the question with aggression or sarcasm?

If you really cannot find anyone to give you feedback, then you can always record yourself, assess your own performance and find ways in which you can improve. It is best if you can video record your presentation, as this will allow you to improve on things like body language and your level of engagement with the audience.

13.8 Summary

In this chapter, I have given you the basic skills on how to become a good presenter, in which you should learn and develop through time. In brief, you will need to:

- Understand the purpose of your presentation and identify the core message you want to relay to your audience.
- Prepare a suitable outline structure and come up with suitable titles/sub-titles for topics/sub-topics.
- Construct your presentation slides (in PowerPoint) on the basis of the outline structure and ensure that points/sub-points have adequately supported the core message.
- Prepare a suitable speech, to accompany the constructed slides.
- Deliver your speech effectively, to influence an ideal outcome.
- Continue to improve your skills as a presenter, by welcoming feedback and remedying any shortcomings for the future.

References

Abrams, R. M., Clapperton, G., & Vallone, J. (2008). *Winning presentation in a day: Get it done right, get it done fast*. Chichester: Capstone.

Bradbury, A. J. (2010). *Successful presentation skills*. London: Kogan Page.

Chalmers, N. (2016). *How to: Give a great presentation*. London: Bluebird.

Davis, M., Davis, K. J., & Dunagan, M. (2012). *Scientific papers and presentations*. Amsterdam: Elsevier/Academic Press.

Dunn, D. M., & Goodnight, L. J. (2016). *Communication: Embracing difference*. New York: Routledge.

Harvey, B. (2008). *Tork & Grunt's guide to great presentations: Arrows not bullets*. London: Marshall Cavendish Business.
Jay, A., & Jay, R. (1999). *Effective presentation: Powerful ways to make your presentations more effective*. London: Prentice Hall.
Ledden, E. (2013). *The presentation book: How to create it, shape it and deliver it!* Harlow: Pearson.
Lowe, D. (2016). *PowerPoint 2016 for dummies*. Hoboken, NJ: John Wiley & Sons, Ltd.
Peel, M. (1998). *Successful presentation in a week*. London: Hodder & Stoughton.
Walker, R. (2015). *Strategic management communication: For leaders*. Boston, MA: Cengage Learning.

Chapter 14
Networking

I have to admit that when I was younger I did not like dealing with people. When I did my PhD., I was happiest when I was working in isolation, designing and conducting my experiments in the laboratory. I dreaded when visitors came around, as I had to explain to them about my project and what I was doing. In the early days of my career, I must admit that I scored zero when it came to networking. I was in complete awe with my peers who were able to network with ease. It seemed as if they have the *gift of the gab*, being so comfortable in the company of others. Undoubtedly, their careers took off at an alarming rate, as they were able to connect with influential people and promote their research (and themselves). As the years went by (like it or not) I had to learn quickly on how to network. The ability for me to network was a necessity, to enable me to stay in research and ultimately help me carve out my own scientific niche. If you are not a natural networker, then do not fret. Like me, you too can learn the art of networking. Hence, the goal of this chapter is to arm you with the right knowledge on how to develop your networking skill.

The chapter is divided into two parts. The first part aims to give you background information, defining what it means to network and how to benefit from the process. The second part will discuss the practicalities, on what you should do in order to be an effective networker. In particular, I will be sharing with you an excellent networking strategy that has been developed by past experts: the guerrilla networking strategy. Finally, I will give you some helpful networking tips, on what to do, should you ever need to deal with office politics.

14.1 What Is Networking?

The term *networking* can be vague and as such it can have different meaning to different people. The traditional act of networking usually involves:

R. Tantra, *A Survival Guide for Research Scientists*,
https://doi.org/10.1007/978-3-030-05435-9_14

1. Attending a meeting.
2. Exchanging business cards.
3. Making small talk.
4. Leaving the meeting with a handful of business cards.

However, this rather basic and traditional form of networking can be ineffective. Quite often, many business cards end up in a sad pile (or in a bin) in many offices all around the world. Hence, take note of your very first lesson: networking is so much more than solely handing out business cards and linking with others. It is also much more than just making your presence known in websites, such as LinkedIn or Facebook. Ultimately, your goal is to be an effective networker, rather than a basic one. Your aim is not only to build new relationships but to maintain your old/current network.

14.2 Why Network?

In order for you to put in the effort needed to network, you will need to understand why it is necessary to network in the first place. Amongst other things, the act of networking will allow you to (Kay, 2010; King, 2008; Singer, 2010):

- Sell yourself (or the organisation that you represent).
- Meet people who will be your potential: allies, investors, collaborators, friends, clients, etc.
- Get advice and Gain knowledge from other experts.
- Gain support, e.g. for a promotion or career change.
- Share your values with like-minded people, so they will eventually want to work with you.
- Influence others, e.g. for the purpose of reaching a consensus.

14.3 Practical Tips: General Do's and Don'ts

A key part to your success as a research scientist lies in your ability to develop an effective networking strategy. You will need to be flexible in your approach, so that you can come up with strategy that works for you e.g. one that supports your personal situation/circumstances. When coming up with a suitable strategy, please take into account the following guidelines (D'Souza, 2010; Grose, 2010):

1. Adopt the right attitude:
   (a) Be positive.
   (b) Do not be selfish.
   (c) Be a good person at heart, e.g. respectful, kind and grateful.
   (d) Do not be pushy or abrasive—otherwise you can end up repelling people.
   (e) Be resilient, e.g. by not taking things personally if someone is being negative or is critical towards you. Forgive, forget and move on.

(f) Show interest in others. Adopt a community spirit and always offer the hand of friendship whenever possible.
(g) Embrace diversity and be curious about other people: Who are they? What do they do? What are their issues/problems? What are their beliefs/value system?

2. Create a good first impression. Work on your personal branding by:
   (a) Looking good (e.g. dress well, take care of personal hygiene).
   (b) Sounding good (e.g. talk clearly in order to articulate your message well). Learn the art of communication, so you can easily build rapport with others. Do not ignore people or their point of views. Listen to others attentively and ask questions when appropriate.
   (c) Adopting a positive body language (e.g. avoid being nervous, angry, bitter or defensive).
3. Identify your networking goal, by knowing what you want to achieve, and then developing a suitable strategy. What outcome do you want to see as a result of your networking activity? For example, if your goal is to develop a relationship with your boss or peers then you might like to invite them for coffee and chat, to get to know them better.
4. Organise and manage your network.
   (a) Grab every opportunity to go to lucrative meetings to expand your network, especially linking yourself with key and important people. Remember to bring your business cards to meetings.
   (b) Use various tools to help you to manage your network. You can create your own database, in order to manage information. In the database, you might like to categorise your network into groups of people, and to link them by common themes or interests. Nowadays, it is common to use the Internet to help manage your network, e.g. via LinkedIn or Facebook.

14.4 Adopting an Effective Strategy: Guerrilla Networking

One strategy that you may like to adopt is the *guerrilla networking technique*, as proposed by Levinson and Mann (2009). The approach (as depicted in Fig. 14.1) is considered superior compared to the more traditional route of networking as it is

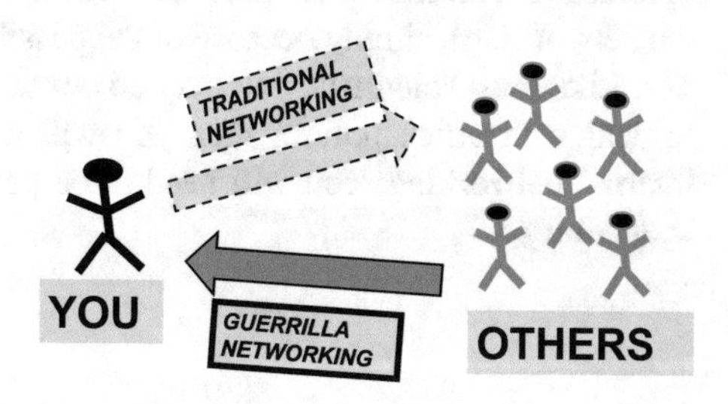

Fig. 14.1 The concept of guerrilla networking is all about drawing people to you rather than the other way around

based on a principle that networking is not about you meeting others but instead creating a situation so that others would want to meet you. Hence, the overall idea is to draw people to you, rather than the other way around.

So, what are some of the methods that you can adopt for guerrilla networking?

In their book, Levinson and Mann (2009) presented 50 different ways on how you can achieve this. In the past, I have used some of their proposed methods, which had resulted in many favourable outcomes, e.g. when I needed to attract research funding or when I had to penetrate into a new scientific community. As a result, my past networking strategy included:

- Becoming an expert (in a particular scientific community): For many years, I was seen as an expert for nanomaterial characterisation within the nanotoxicology community.
- The use of appropriate publishing channels in order to promote myself and my science. I wrote books and peer-review publications often with other project partners. The collaborations with others were particularly important, in order to link my name to influential people in the community.
- Saying thank you a lot and acknowledging people in my publications.
- Being courteous and polite to others.
- Being of use to others. I happily gave free advice to people.
- Offering the hands of friendship whenever possible: Remember that given the chance, people will want to do business with their friends or those they can trust.
- Striving to be unique: As such, I always express my opinions and stick to my beliefs, even when I am sometimes in the minority. Whenever I felt that I was not being heard, I decided to write strong opinion piece articles, which were published in high-impact journals such as Nature and Lab-on-Chip. These papers acted as promotional material and a useful starting point to link myself to other like-minded people.

14.5 Internal Networking and Office Politics

When we think of the word *networking*, we often associate this with external networking, i.e. when you are expected to go out into the world, represent your company and network for some purpose, e.g. getting funding, joining a consortium and promoting a product. However, networking can also be internal, i.e. within your organisation. At times, internal networking can be more of a challenge (than external networking), simply because of the possibility of having to deal with office politics. Like most scientists, you may be out of your depth here. However, if you need to deal with office politics then you will need to dedicate some of your time to internal networking. You will need to be people savvy if you want to survive and keep your job.

Before going any further with this topic, I will need to clarify as to what I mean by office politics. Office politics is all about the power within an organisation and how you (or others) are able to acquire (some of) that power.

So, how can networking be useful in an environment riddled with office politics?

Quite simply, by knowing where the power is, you can then use your networking skills and develop the right strategy to help you tap into this power. As such, your ultimate networking goal is to strengthen your position within the hierarchy of who's who amongst the pecking order within your organisation.

For you to network effectively in the backdrop of office politics, you will first need to understand the cause/source of the office politics. Office politics can stem from (James, 2014):

- The organisational structure and culture: It is possible that your organisation will want to promote a bit of (healthy) office politics, as a way to initiate competition between colleagues. The aim here is to propel people to work harder in order to survive in a competitive environment.
- Personal ambitions and desires, with individuals wanting to secure company rewards, promotion, research funding, etc.

So, why should you take notice of office politics?

If you work in a particularly politically charged environment, you simply cannot choose to ignore it. Doing so can have negative impact, potentially making you more vulnerable. As a result you may (James, 2014):

- End up feeling isolated, whilst your colleagues scam and scheme to think of ways to get rid of you or your technical area.
- End up being backstabbed by colleagues, who will not think twice before taking away your ideas and claiming them as their own.
- Be subjected to colleagues who will play tricks on you, e.g. making you a scape-goat for something that was not your fault.
- Not get the proper recognition for the good things that you do or have achieved.
- Get unfair appraisals, in which your boss criticises you unfairly and constantly.
- Be constantly put under the magnifying glass, in which others go out of their way to criticise your work continuously and hinder your progress without justification, in order to destroy your credibility as a scientist.
- Have to deal with aggressive people or ruthless behaviour, without real logic as to why they might be doing this.

Although I do not recommend that you play dirty office politics, it is vital however that you are aware of it (of its existence) and to devise a networking strategy in order to strengthen your own position and to minimise its negative effects. When developing a suitable strategy for your own particular situation, please consider the following tips (especially useful when you are a new starter in a company):

1. Map out your internal network by studying the people in your organisation (Clarke, 2012). What interests them? Understand the hierarchy of that company, appreciate its dynamics and identify key players, such as those who are seen by management as the rising stars (as opposed to those who are becoming less popular). Note that the key players may not be the senior people but the people who can for whatever reason (e.g. due to technical skill/knowledge) influence how decisions are made.
2. Identify your own goals and aspirations. You will need to be clear on what is important to you and where you want to head within the organisation. Who do you want to work with? Who do you want to avoid? Are you looking for a promotion?
3. Continuously, be aware of your own position within the hierarchy of power. Is your position of strength or weakness? Is there a way that you can strengthen your position further?
4. Be sensitive to your surroundings. Be on your guard and learn to read between the lines. Look at what is being said and how it is being said.
5. Identify the *bad apples*, especially those who are also influential (Furnham & Taylor, 2011). Remember they can be sly and cunning. They may not be so brash and blatant in their dealings. They are able to change their image easily depending on the situation, just like a chameleon. However, sooner or later you will learn of their real character. It's best not to be too close with such people. If you do need to work with them, then do not antagonise them. Like the warning labels on chemicals, remember that they are toxic … so handle with care!
6. Realise that not everything is within your control. For example, you might find yourself to be a victim in a very bad situation e.g. in which colleagues are trying to conspire against you. If you are ever in this position then you will need to (Kay & Institution of Engineering and Technology, 2009):

   (a) Identify those people who are conspiring against you. Who are they? What are their motives? By understanding the situation, you may be able to accept it better, which will help you to perceive the situation differently. Remember that often office politics is not meant to be a personal attack.
   (b) Maintain a professional outlook. Develop a thick skin—ignore criticism, don't be emotional or take things personally. Do not allow anyone to intimidate you. Do not take revenge and participate in playing dirty office politics yourself. Do not be confrontational or show offence. Do not show sign of inferiority. Smile whenever possible.
   (c) Choose to address the situation and not ignore it. If you shove things under the carpet, then you will risk being left on the sidelines. Find out more information by touching base with your internal network. You will need to get the facts, i.e. as to what (and why) it is happening.
   (d) Stay calm and think logically/strategically at all time, to develop a networking strategy that will work for you and your particular situation.

7. Remember that everyone in the organisation is important and thus it is vital that you bond with ALL colleagues in the early phases of your career. When trying to

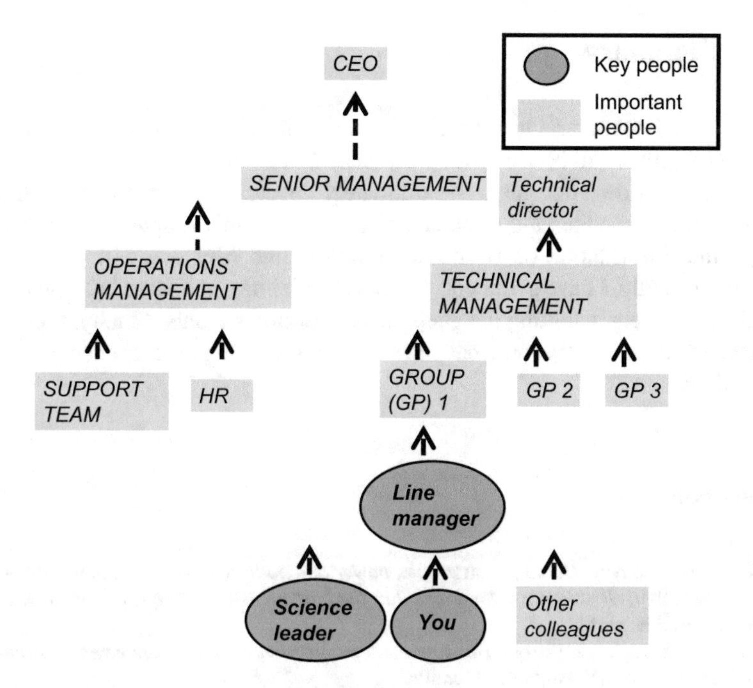

Fig. 14.2 Illustration depicting your relationship with others in the organisation. Although everyone is important, you will need to concentrate on certain key people (circles)

bond with others, remember to brush up on your communications skills, e.g. listening attentively, as well as to asking appropriate questions to make the other person feel important/special. If possible, get to know everybody and aim to get their support. Whilst it is true that you should treat everyone with respect, your focus should be on strengthening your relationship with the key players. Figure 14.2 aims to illustrate this point. As shown in the figure, besides yourself, the key players here are your line manager and the science leader who is responsible for ensuring that the science quality within your group is maintained at a certain level. Although he/she may not be at the top end of the hierarchy, he/she will ultimately influence key decision-makers, e.g. senior management such as technical director and CEO.

8. Should office politics jeopardise your job situation in the company, then you will need to identify immediate action plans to minimise the damage. Can you rely on your internal and external support network for support or advice? Can you be flexible in your approach in order to find a solution? Is there a job opportunity elsewhere, either internally or externally? Remember that it is often easier to change yourself and your plans rather than expecting others to change. Try to visit HR and colleagues in other departments to see if there are any upcoming internal opportunities not yet advertised. If you are looking for a new job, then my advice to you is to do this whilst you still have one. Remember that it will become more difficult for you to get another job once you become unemployed. Also, consider the possibility of self-employment, if all else fails.

14.6 Summary

Most of us are not natural networkers. Yet, the skill to be able to network effectively is a necessity that will benefit your professional life. An important aspect when you network is to know your goals and ultimately develop a suitable strategy that will help you to achieve the kind of outcome you want. In this chapter, I have presented background information on what networking is and why you will want to do it. More importantly, I have given you practical guidelines on how to become an effective networker, e.g. adopting the guerrilla networking strategy. Finally, I have given you some networking tips on what you should do, should you ever need to deal with office politics.

References

Clarke, J. (2012). *Savvy: Dealing with people, power and politics at work*. London: Kogan Page.

D'Souza, S. (2010). *Brilliant networking: What the best networkers know, do and say*. Harlow: Prentice Hall Business.

Furnham, A., & Taylor, J. (2011). *Bad apples: Identify, prevent & manage negative behavior at work*. New York, NY: Palgrave Macmillan.

Grose, R. (2010). *How to sell yourself*. London: Kogan Page.

James, O. (2014). *Office politics: How to thrive in a world of lying, backstabbing and dirty tricks*. London: Vermilion.

Kay, F. (2010). *Successful networking: How to build new networks for career or company progression*. London: Kogan Page.

Kay, F., & Institution of Engineering and Technology. (2009). *How to build successful business relationships*. Stevenage: Institution of Engineering and Technology.

King, N. (2008). *Networking: Work your contacts to supercharge your career*. Oxford: Infinite Ideas.

Levinson, J. C., & Mann, M. (2009). *Guerrilla networking: A proven battle plan to attract the very people you want to meet*. Bloomington: Morgan James Publishing.

Singer, T. (2010). *Networking unplugged*. New Year Pub.

Chapter 15
Dealing with Difficult People

In an ideal world, you will be in a position to choose who to work for/with. Preferably, you will want to have an understanding boss, colleagues who are your friends, dependable technical staff, etc. However, the reality can be very different. You can end up with a boss who belittles you, colleagues who want to sabotage your research credibility and technical staff with an attitude problem. They are your so called difficult people. Even though you may not like to deal with these people, quite often your research (and career) can depend on them. Remember that no matter how annoying your difficult person may be, he/she can potentially be the gatekeeper to your future success. For example, you may need a bright technical staff to build you an essential piece of kit for an experiment that will eventually lead to innovation and glorious times ahead.

A key aspect in dealing with a difficult person is to have empathy. In order to do this, you must identify where the bad behaviour is coming from, thus understanding his/her intent. As such, in the first part of this chapter, I will discuss some common intent types behind human behaviour. By understanding intent, you will be able to devise a suitable strategy that will help you deal with your difficult person. Note that most of the information on human intent (and what to do) presented in this chapter has been based on work previously published by Brinkman and Kirschner (2002). No doubt, there are different theories out there that can explain human intent and behaviour, e.g. Maslow's hierarchy of expectations (Pichère, Cadiat, & Probert, 2015), transactional analysis (Lapworth & Sills, 2006) and Myers-Briggs behavioural type (Briggs & Myers, 2010). However, I have chosen to focus on the model of human behaviour as described by Brinkman and Kirschner, simply because of their practical and effective approach when dealing with difficult people. The second part of the chapter will deal with the topic of conflict and what to do in a conflict situation. The last part of the chapter aims to understand a special kind of difficult person: an office bully. I will give you background information on bullying, as well as practical tips on what to do if you ever find yourself being bullied by a boss or colleague.

R. Tantra, *A Survival Guide for Research Scientists*,
https://doi.org/10.1007/978-3-030-05435-9_15

15.1 Understanding Your Difficult Person

Although dealing with a difficult person can be challenging, there are certain steps that you can take for the purpose of damage control. First and foremost, you will need to have empathy. This is important so that you can accurately identify the intent of your difficult person. According to past experts (Brinkman & Kirschner, 2002), there are four intent types that you should be familiar with (as depicted in Fig. 15.1). Your problem person may be acting difficult because he/she has the intent to:

- Get things done.
- Do things right.
- Want to get along (with others).
- Be appreciated (by others).

You may argue that as humans, we all will want to get things done, get things right, get along with others and be appreciated. Who doesn't? Whilst to some extent this is true, you must appreciate that your difficult person will have a much stronger association with one of the intent types compared to others, particularly if he/she is under stress.

15.2 Identifying Intent to Develop a Strategy

According to Drs. Brinkman and Kirschner (2002), a person with a particular intent type will exhibit certain classic behaviours. By identifying what these are, you will then be able to accept the situation and alter your own attitude/behaviour towards him/her.

Fig. 15.1 Identifying the intent of your difficult person

15.2.1 If Someone Wants to Get Things Done

When a problem person has an inkling that things may not get done, his/her intention is to do whatever it takes to move things along. Thus, his/her focus is on achieving the final results. As such, he/she can display classic behaviour of being:

- Controlling, often being outspoken and assertive
- Aggressive or abrasive
- Rude, possibly making sarcastic comments or irrelevant remarks
- Arrogant, to display his/her knowledge/expertise

In order to deal with this kind of problem person, you will need to alter your own attitude and behaviour in order to cope. Specifically, you will need to:

- Stick to the main points/facts and not waffle. Be assertive (not show any sign of weakness) when delivering your message.
- Have the courage to stand up for yourself and speak up if your difficult person is out of line. The aim here is to expose any unwanted behaviour, for example if your problem person makes rude comments. If you think your problem person is wrong then say so, but make sure that you:
  - Are 100% certain that your difficult person is indeed wrong. Never engage in a confrontation if you are uncertain.
  - Never bluff your way out of a situation.
  - Own up and apologise (only once) if you have made a mistake.
- Tread carefully if your problem person becomes arrogant. He/she is only doing this in order to display his/her immense knowledge. It is best not to confront such a person head on, as this type of person does not like his/her authority being questioned. Hence, your best approach is to:
  - Listen attentively.
  - Not dismiss his/her ideas immediately.
  - Be indirect on how you introduce new ideas into the conversation.

15.2.2 If Someone Wants to Do the Right Thing

When this problem person has an inkling that things are not being done right, he/she then displays the classic behaviour of:

- Being a perfectionist or
- Being a whiner or
- Saying no to everything or
- Say nothing at all.

When you are confronted by such behaviours, you will need to alter your own attitude and behaviour.

- If he/she whines, you must maintain a positive attitude. You need to see if any of the complaints are genuine and if there is indeed scope to improve things. Realise also that sometimes people tend to whine as a way to release pent-up anxiety.
- Do not agree or disagree with whiners. Be practical in your approach, so that you can come up with potential solutions, rather than magnifying the problem.
- If you have a perfectionist on your hands then he/she often displays a negative outlook. If this is the case, then do not attempt to try and change his/her mindset straightaway. You will need to give this type of person time to think things through. Your tactic is to give practical advice. However, you must first acknowledge that you have understood the negative outlook. You then need to show him/her the future, i.e. providing a vision on what can happen if his/her negative outlook persists, e.g. it may result in a project not moving forward.
- If your difficult person simply does not make an effort to talk, then your best bet is to be patient. Your goal here is to break the ice. You can do this by asking the person open-ended questions … questions that cannot be answered with either a yes or a no. You can also introduce humour into the conversation in order to break down the person's armour against his/her unwillingness to talk.

15.2.3 *If Someone Needs to Get Along*

When this problem person has an inkling that he/she is not getting along with others, then he/she can display several classic behaviours in order to please others, to include:

- The need to seek approval from others.
- Over-committing, taking on far too much work: He/she will end up saying yes to everything and often not being able to deliver.

In order to deal with this type of problem person, you will need to remind yourself that he/she is genuinely trying to be nice. As such, you will need to alter your own attitude and behaviour to deal with such a person. Here are some practical tips for you to try:

- Do not be rude or show that you are upset with the problem person. Instead, create a friendly environment (say sorry if it helps), to discuss issues (even if you feel like throttling them!). This kind of person will need reassurance more than anything.
- Do not push your problem person into a decision, if he/she is indecisive. You will need to back off if you feel that the tension is mounting and approach him/her on a different day, so that you can discuss things when the environment is more relaxed.
- Give practical advice, especially if he/she starts procrastinating. This type of person needs support more than anything else, in order to reduce the feeling of

being overwhelmed. For example, you can help him/her to plan for things, to ensure his/her commitment. You can also explore different options together and then write pros and cons, to identify future actions.

15.2.4 If Someone Wants to Feel Appreciated

When this type of problem person has an inkling of not being appreciated, then he/she will display attention-seeking behaviour, for example by:

- Becoming arrogant, talking about stuff that he/she really knows little of.
- Becoming explosive.
- Making snide remarks at your expense, so that he/she can feel more superior.

In order to deal with this type of problem person, you will need to alter your own attitude and behaviour, by:

- Asking probing questions, especially if you are dealing with someone who is arrogant. By doing so, you will be revealing to others that your difficult person really does not know his/her material. Once you have discredited this person, you must resist the temptation to isolate or make fun of him/her. Instead, extend the hand of friendship.
- Staying calm, if your difficult person becomes explosive: When this happens, your goal is to reduce the intensity of the explosion. For example, you can show compassion and aim for the heart by reassuring that you have understood his/her concerns and work on from there. You will need to do whatever it takes to calm the situation down. The worst thing that you can do to this person is to lose control and get angry yourself.

15.3 General Advice: How to Deal with a Conflict

Although the above guidelines given by Drs. Brinkman and Kirschner aim to neatly categorise people and on how to deal with a difficult person, you must understand that human beings (and the situations in which you find yourself in) can be extremely complex. As such, you will need to treat every conflict with your difficult person as unique. By all means, refer to the above guidelines for consideration, but remember to be flexible in your approach and do not be afraid to modify what you need to do, until you find something that works.

If you are ever in a conflict situation, you will need to consider the following general dos and don'ts:

- Change your attitude and how you perceive conflicts in general. Remember that:
  - You will face some kind of conflict sooner or later, as it is an inevitable part of working life.

- You will need to embrace diversity, by acknowledging that people are different and thus will respond to situations differently.
- When addressing an issue, you will need to separate the issue from the person. Do not perceive the conflict as a personal matter.
- Escalating a conflict will not help. You must try to diffuse the situation and must remain calm at all times.

- Find out the nature and source of the conflict before trying to come up with potential solutions. You can do this by (Cloke & Goldsmith, 2011):
  - Listening attentively so that you fully understand the nature of the conflict. If you cannot understand another person's point of view, then say so and enquire on the exact nature of the problem. Ask for specific details so that you can develop the bigger picture. If you are missing any information, then take time out so that you can investigate further.
  - Being sensitive to your environment: For example, it may be that you are being gossiped about by others, in which case you will need to nip this in the bud, e.g. by going to the person who is gossiping about you, stating your concerns and correcting any information that is not true.
- Identify practical solutions, to move forward. For example (Cloke & Goldsmith, 2011):
  - If you are wrong, then you need to apologise and say how you can do things differently next time around or how you are going to remedy the situation.
  - If you are right, then stand up for yourself. However, remember to present your argument well, by sticking to the facts, reminding your difficult person of the bigger picture, e.g. what the acceptable standards are.
  - Try to find some common ground between you and your difficult person. Then, find ways in which you can work with one another to reach a common goal.
  - If the issue persists, then your next step is to seek internal or external support. For example, if you are a boss then you can go to human resources, not only for advice but also to make sure that you are in line with employment law.

Whatever solutions you come up with in order to resolve a conflict, remember that you will need to clearly communicate any solutions proposed. If necessary, write them down in an e-mail and send to those who are involved in the conflict.

15.4 Dealing with Bullies

A bully is a different kettle of fish from your average difficult person. With a difficult person, you can more or less manage, if you have accurately identified his/her intent and adopt a suitable strategy afterwards. Things are not so simple with bullies, as they may have a personality disorder or exhibit psychopathic tendencies.

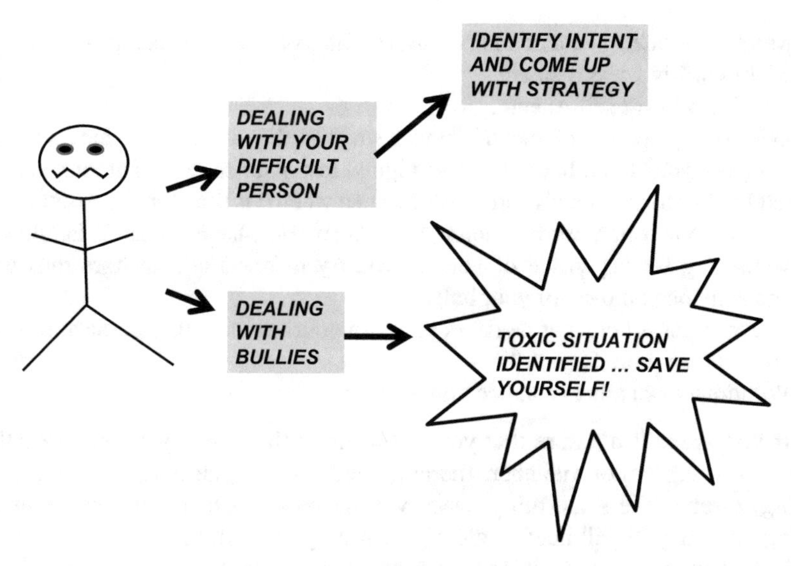

Fig. 15.2 Bullies are different from your difficult person

Potentially, you can be in a toxic environment. Hence, your goal is to save yourself (as shown in Fig. 15.2).

When you first encounter a bully, your first inclination is to ask yourself the following question:

Was it my fault?

Quite simply—the answer is a straightforward:

NO

After this, you may try hard to understand as to where the bad behaviour is coming from. Perhaps your bully had a challenging childhood. Your efforts hard to empathise somehow does not help matter much. My first advice to you is this: when you are confronted with a bully, it is not your job to figure out why, as you are neither a psychologist nor a psychiatrist. Your main goal is to identify the options and to eventually save yourself! But before delving into what options are available, you will need to be certain that you are in fact being bullied.

According to experts, there are some common patterns that are associated with bullying. A bully often shows two classic behaviours (Clifford, 2006; Field, 2010):

1. Bullies will show a clear effort to cause you harm. This can take on different forms such as insults, humiliation, spreading gossips, belittling, shouting, dishing out unfair criticisms, removing suitable work from you, giving unreasonable targets/deadlines and ignoring your opinions.
2. Bullies display continual negative behaviour, i.e. his/her bad behaviour is not a one-off. It is persistent, often becoming worse through time.

You know if you are being bullied, as the effects of bullying can make you (Clifford, 2006):

- More emotional (you tend to cry a lot and be more nervous than usual).
- More sensitive, less resilient.
- Develop a lower self-esteem.
- Develop symptoms of mental illness such as depression, in which you stop enjoying your food, have sleepless nights, lose your sense of humour, etc. You tend to live for weekends and days off. Then when it is time for you to come back to work, you often become depressed and anxious, having that all familiar gut wrenching feeling inside of you, as you try to brave another torturous week ahead in the company of your bully.
- Become less efficient at work, as you lose your self confidence, make mistakes and become more forgetful.
- Withdraw from others and become isolated.

If you are still not sure that you are being bullied, then you need to talk to somebody, e.g. union member, therapist and a colleague who has gone (or is going) through the same thing. Once you have established that you are indeed being bullied, you will need to clearly identify your options.

Before making any decisions, you must be at a stage in which you have discussed it is wise to discuss your situation with a confident, firstly to clarify the situation and seek emotional support. This can be a friend, spouse or a close colleague. Never navigate the problem by yourself and do not put your head under the sand hoping that the problem will go away. You must strive to take action early, to regain control. You must understand what choices you have and decide on how you want to proceed. Here is a list of some possible practical things to consider (Field, 1996):

- Start a diary and log things down. Write down events, dates, what the bully did to you, how you felt at the time and if witnesses had been present. If there are were witnesses, then get them to write statements (if they are happy to do so). The need to log events associated with the bullying is important, not only for your own sanity but also if you were to take the matter further.
- Before taking the matter via legal proceedings, you will need to consider if this route is indeed right for you. You will need to talk to a professional, e.g. lawyer, union member and citizen's advice bureau, to not only clarify your position but also establish if you indeed have a case. However, in the case in which a criminal offence has been committed, e.g. if you have been sexually attacked, then you must report the matter straight to a higher authority, such as the police.
- Consider seeking internal help to voice your grievance. Investigate what your company's policy is on bullying. Theoretically, you can report the bully to your boss or your boss's boss or HR. However, it is wise to assess your position carefully to have an idea if your complaint will be taken seriously. For example, it may be that a company has certain policies against bullying but it does not mean that they will implement and act tough on the bullies themselves. The bully may only get a slap on the wrist, as a result. Then, where does this leave you? Be wary of agreeing to meetings with HR. You need to ask yourself what can HR do? This is especially true if your case is weak, e.g. you have no reliable witnesses to confirm the bullying.

- Consider finding another position within the company or look for a job elsewhere. From my point of view, this is the best option to overcome bullying (particularly if you are dealing with a deranged lunatic).
- Consider the option of resigning. You should only consider this option if the bullying is particularly bad and you see no other way out.

Quite often when you are being bullied, you become emotionally entangled that you start believing in what your bully is saying to you and that you begin to accept how things are. Never let this happen, as you must strive to regain control and tackle the problem head on. Overall, think logically and practically, to decide what route you will want to take, which should be right for you.

Remember that whatever happens, you must take care of yourself (physically and mentally). Make sure that you eat well and do some exercise to reduce your stress level (refer to Chap. 1). If you are suffering from any physical or mental ailments as a result of the bullying, then you will need to talk to the relevant medical profession.

15.5 Summary

The world of work can be a tough place. As we do not have the luxury to pick our colleagues, boss, etc., it is inevitable that we will need to deal with troublesome people at some point in our lives. Based on the work of past experts, I have given you practical guidelines on how to deal with difficult people. In summary, you will need to:

1. Empathise and identify intent.
2. Change your own attitude/behaviour.
3. Develop and adopt a suitable strategy.
4. Realise that the nature of any conflict is unique. As such, your first attempt to deal with the conflict may not work. It is imperative that you adopt a flexible approach on how you want to proceed, until you find a formula that works.

Finally, the chapter ends on the topic of bullying. If you are being bullied, it is vital that you do not ignore the symptoms of bullying. Make it your mission to take actions early on. The chapter presents some practical action points to consider if you are being bullied. Remember that there is no right or wrong way to deal with a bully, as the steps you take will be dependent on your personality and personal circumstances.

References

Briggs, I., & Myers, P. B. (2010). *Gifts differing: understanding personality type – The original book behind the Myers-Briggs Type Indicator (MBTI) test eBook: Isabel Briggs Myers, Peter B. Myers: Amazon.co.uk: Kindle Store*. London: Davies-Black.

Brinkman, R., & Kirschner, R. (2002). *Dealing with people you can't stand: How to bring out the best in people at their worst*. New York: McGraw-Hill.

Clifford, L. (2006). *Survive bullying at work: How to stand up for yourself and take control*. London: A. & C. Black.

Cloke, K., & Goldsmith, J. (2011). *Resolving conflicts at work: Ten strategies for everyone on the job*. San Francisco: Jossey-Bass.

Field, E. M. (2010). *Bully blocking at work: A self-help guide for employees and managers*. Bowen Hills, QLD: Australian Academic Press.

Field, T. (1996). *Bully in sight: How to predict, resist, challenge and combat workplace bullying: Overcoming the silence and denial by which abuse thrives*. Wantage: Success Unlimited.

Lapworth, P., & Sills, C. (2006). *An introduction transactional analysis: Helping people change*. London: Sage Publications.

Pichère, P., Cadiat, A.-C., & Probert, C. (2015). *Maslow's hierarchy of needs*. Namur: 50Minutes. com.

Chapter 16
Project Management

As a research scientist, you will need to dedicate some of your time to acquire soft and transferrable skills so that you can learn to do other things besides technical research. This not only is important for you to grow as a person, but will also you give you more options and make you more employable, when you are looking for a (permanent) job. One of the most useful things that you can do is to learn how to manage a project. By acquiring the skill set necessary to be a project manager, you will be in a position to ensure that any project/s (in which you may be technically responsible for) will be successfully delivered. In addition, you may in the future decide to have a career change, to work away from the bench in order to become a science project manager full time.

The goal of this chapter is to give you a quick overview of project management and how you can develop the right skill set to be a good project manager. I will start the chapter by giving you some background information on:

1. What a project is.
2. What project management is all about, specifically focusing on the different phases associated with a project management's life cycle, namely the launching, planning, execution and closure phases.

I appreciate that there is so much more useful background information on this topic, more than what I am planning to cover in this chapter. Hence, if you want to learn more, then I suggest that you refer to suitable past references (Baguley, 2008; Cooke & Tate, 2011). There is also the option to take a formal management course, such as Prince2 (Projects In Controlled Environments) (Barker & Cole, 2009). Going on such a course will not only give you more of a structured framework of the different methods associated with project management but upon successful completion you will also receive a certificate of qualification (Mathis, 2014). However, from my own experience of working as a project manager and also being managed by others, having a formal qualification does not guarantee that you will be good at your job. Quite often, many project managers lack the three vital skills, namely:

R. Tantra, *A Survival Guide for Research Scientists*,
https://doi.org/10.1007/978-3-030-05435-9_16

- How to handle people.
- How to solve problems.
- How to manage time.

As such, my focus for the remainder of the chapter will be based on these three sub-topics.

16.1 What Is a Project?

A project is a temporary endeavour in order to create a unique product, services or processes to be delivered within a set time frame (Baguley, 2008; Cooke & Tate, 2011). The operative word here is unique, which means that the outcome will be created for the first time; in this sense, a project differs from operations (in that operations use established methods and tools).

16.2 What Is Project Management?

Project management is a set of processes, which is necessary to deliver a project. Hence, as a project manager, you are expected to carry out a sequence of activities to be conducted over a limited time. According to past experts, there are four phases associated with project management that you should be familiar with, namely (Cooke & Tate, 2011; Young, 2006):

1. The initiation phase: this is conducted at the start of the project.

   In this phase, you will need to clarify the requirements of the project and to identify priorities. You will need to:

   (a) Ensure that there are sufficient resources (in relation to the financial budget, accessibility of capital equipment, accessibility to suitable team members, etc.) to help with the successful delivery of the project.
   (b) Ensure that the project is sound by clarifying the approach, strategy, timeline and scope of work.
   (c) Identify mitigation actions to reduce any potential risks to the project.
   (d) Identify benefits versus costs.
   (e) Get approval from senior management to proceed.

2. The planning phase (to be conducted after gaining the approval from senior management for the go-ahead): During this phase, you will need to:

   (a) Clarify the project and your future plans for the project. You should identify both high-level and detailed-level planning to team members. You will need to ensure that team members understand the context, structure, strategy and process.

(b) Establish the necessary commitment from team members, to the plan and delivery time. Any foreseen discrepancies should be clearly identified at this stage and any issues to be resolved as soon as possible.

3. The execution phase: This is the stage in which team members get involved in the necessary activities to successfully deliver the work. You will need to deal with any problems that (may) arise. You will need to keep the project moving forward whilst at the same time ensure that your customers will be satisfied with the outcome. If in doubt, you will need to converse directly with your customers, to detail any changes or deviations to be made to the original plan.
4. The closure phase: This stage is all about the need to finalise the project. As such:

(a) Formal reports will need to be filed, e.g. to upper management and customers. Ensure that any lessons learnt from the project have been disclosed.
(b) All deliverables will need to be sent to the customers. You will need to ensure that an acceptable level of customer satisfaction has been achieved. You can do this by sending out feedback forms to customers.

16.3 How to Manage People

Key to being a good project manager is to have the necessary skill set to manage people, especially those who will be working directly on the project. Past experts have identified some of the common traits that you will need to develop in order to manage people well, namely the need to have (Gilman, 2013; MacDonald & Tanner, 2012):

- Good emotional intelligence, i.e. the ability for you to be sensitive to the needs of the environment and others (particularly your customers and team members). By being more sensitive to what others need, you will be able to anticipate problems and issues before they arise and provide potential solutions in advance.
- Excellent communication skills: Remember to listen attentively (in addition to speaking). Be patient and consider all of the suggestions given to you but at the same time make sure that everything is still aligned with the goals of the project.
- Self-awareness, thus knowing your own strengths and weaknesses: For example, in order to manage a technical project, you will need to ensure that you acquire sufficient technical knowledge. This is important so that you can understand what team members are doing and why they are doing it. This will also allow you to appreciate the methods and tools used by your team members in order to deliver the project. You can acquire the right technical necessary knowledge by doing background reading talking to team members.
- Good networking skills, which is your ability to build new relationships and maintain existing ones. For more information on networking, please refer to Chap. 14.

- Good leadership skill: This will allow you to build your team and identify key personnel, i.e. those who you can count on to deliver the project successfully. In order to lead effectively, you will need to have a positive attitude whilst having clear alignment with the project's mission. You will need to promote team cooperation and raise morale. For more information on how to lead a team, please refer to Chap. 17.

16.4 Problem-Solving

At times, being a project manager is not a lot of fun.

Why?

Because you will be faced with problems, problems and more problems. Hence, a vital skill to acquire is in your ability to solve problems and make informed decisions. This is crucial, as a problem can become an obstacle or a roadblock that can hinder the ability for your team members to achieve their goals/targets (as required by the project).

So, if faced with a problem, how do you go about solving it?

It is difficult for me to give you an exact formula as every problem is unique. For a start, the size and severity of a problem can vary. You may be faced with a small problem that is relatively easy to solve or one that has little impact on the project. On the other hand, you may need to solve a big problem, which is not only complex in nature but can also have a significant (negative) impact on the project. Whatever the problem may be, remember that as a project manager, all problems will demand your immediate attention, so that the project can move forward. In fact, an excellent project manager will try whenever possible to anticipate the problem and have the corrective actions in advance (well in advance.).

When trying to solve a problem, there are some tips that can help (Baguley, 2008):

1. Keep yourself cool. Do not stress out and never take things personally.
2. Define the situation and focus at the root cause, in order to identify possible actions (rather than focusing too much on the symptoms of the problem). As such, you must collect all of the relevant information (from your team members or external sources) in order to understand the situation better. Some potential root causes can include (Manser, 2012):
   (a) Having goals/aims of the project not clearly defined, resulting in activities that are poorly aligned: This may be due to poor communication (possibly from you to begin with).
   (b) Unrealistic expectations of project aims. For example, you may discover that there are insufficient resources, e.g. due to budgeting constraints.
   (c) Poorly defined roles of the individual team members: This may be due to a lack of ownership/accountabilities for particular tasks. It may also be that certain team members do not have the adequate experience/skill set to deliver.

   (d) The team simply not gelling together: Poor teamwork efforts may be due to the fact that some members may feel unappreciated or unmotivated.
   (e) Weak leadership, from either yourself or other (e.g. technical) leaders.
   (f) The lack of monitoring on progress and performance of team members.

3. Define your objectives. Ask yourself: What outcomes do you want to see by solving the problem?
4. Evaluate what options you have and weigh out the limitations versus benefits arising from any planned actions, so that you can decide what actions to take. Overall, it is best to go for an option that is the simplest, most efficient and economical. If you are unsure as to what you should do, then opt for a trial run to see what happens. At the end of the trial run, you can choose to:

   (a) Continue with the existing corrective action, or
   (b) Reject this and opt for a different option, or
   (c) Modify your chosen action.

At this point, I would like to expand on point (2) above. I want to make you aware of some tools that you can use, to help you find the root cause of a particular problem:

- Brainstorm methods (Gido, Clements, & Baker, 2017). Brainstorming is to generate lots of ideas, useful if you want to come up with a creative solution to a problem. Although you can brainstorm for ideas by yourself, it is best to conduct this on a group level. If you are in a group setting, then a chairperson and note-taker must be appointed. It is the responsibility of the chairperson to make the brainstorming rules clear to others, primarily that all ideas are welcomed and that you cannot criticise other people's ideas. The chairperson will also need to create an honest and open environment to encourage participation. The note-taker will need to record people's ideas and put them on flip chart or white board. He/she cannot modify or edit a person's idea. At the end of the brainstorming session, the group will then need to make sense of the ideas generated by:

  – Categorising the ideas, i.e. merging similar ideas together.
  – Cross-fertilising ideas, i.e. amalgamating them together.

The group will then need to identify which of the shortlisted ideas are the most promising. Subsequently, suitable action points for implementation can then be proposed.

- Draw a fish bone diagram (also called the Ishikawa or cause and effect diagram) (Pries & Quigley, 2012). As illustrated in Fig. 16.1, this is a visualisation tool to identify potential causes of the problem, which can fall into one of several categories, such as equipment, process, people, method, environment and management.
- Use the what, where, when and who method. As the name implies, the method aims to clarify what is happening, then where, when and who is involved.

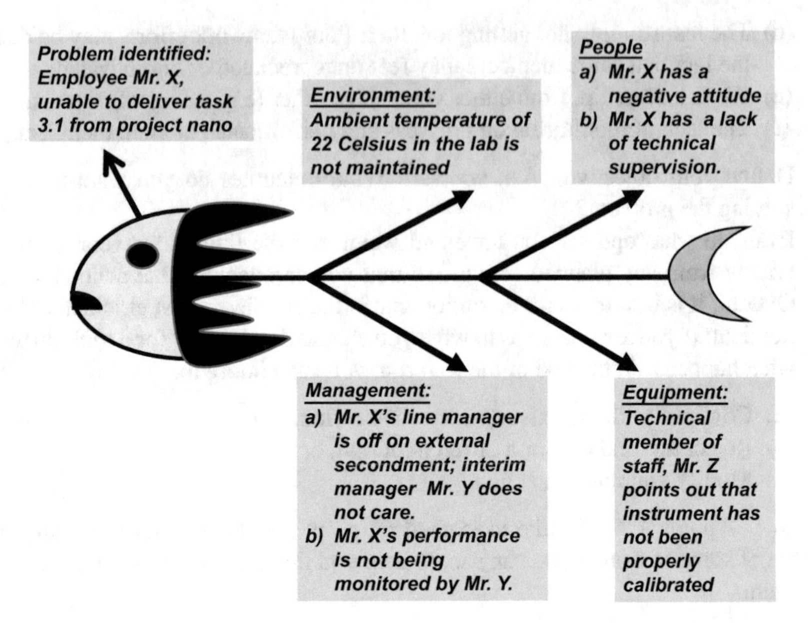

Fig. 16.1 An illustration of a fish bone diagram to show possible cause and effect of an identified problem

At this point, I would like to iterate that you will need to appreciate the uniqueness of every problem that comes your way. As such, always adopt a flexible approach to find a solution that works.

To illustrate the uniqueness of every problem and how you propose to solve it, please refer to Table 16.1 summarises two (hypothetical) case studies that aims to illustrate this point. Although both case studies deal with staff issues, they differ in that one deals with a group conflict and the other deals with an underperforming staff. As such the potential strategy on how a good project manager will deal with the each situation will also need to be different, as shown in the table (Armstrong, 2013).

16.5 Time Management

As a project manager, you will need to have excellent time management skill. If you cannot manage your own time, then you will find your job overwhelming as you will fail to complete tasks.

The key to time management is to develop a plan of proposed activities, in order to improve productivity. Hence, consider:

1. Having a calendar to organise activities and manage time: For example, you can use a calendar through Microsoft Outlook or use a relevant app. In the calendar you can block time schedules for various activities, e.g. meetings and personal appointments. You also have the option to share your calendar online with others.

Table 16.1 Potential strategies on handling two very different types of staffing issues

Problem situation	Potential strategy to handle the situation
Dealing with a conflict between people	Set up meetings, first with individuals and then with everyone else involved. At each meeting, listen attentively and ask open-ended questions to start off with, in order to collect information and filter out the facts. Recognise that feelings can be exaggerated but you must allow them to be expressed. Act as a counsellor in order to mediate the situation. Encourage individuals to actively come up with their own solutions—rather than giving orders. Get commitment from all concerned on the proposed actions. Monitor the situation. Involve another senior member of staff, e.g. science head, as a last resort.
Handling an underperforming staff	Have a meeting with the individual concern. Listen attentively and collect facts from your individual. Understand the source of the problem, i.e. why he/she is underperforming. Does he/she lack certain skills and thus require further training? Does he/she have a habit of taking on too many tasks and thus overpromising in order to please others? Does he/she tend to procrastinate? If so, can you pair the individual with another member of staff, in order to help the matter? Discuss possible options to identify action plans together. Move forward by getting commitment from the individual. Give him/her time to improve performance. If not successful then you will need to issue a verbal warning. Take disciplinary action as a last resort and involve HR to make sure that you are within employment law.

2. Creating a Word document detailing a list of activities for work only: Please refer to the "My day-to-day task" example below, to illustrate. In this document, you can write down all of the tasks that you will need to do. Ensure that tasks are written as a SMART deliverable, i.e. *S*pecific, *M*easurable (in which you are able to measure the desired outcome), *A*chievable, *R*ealistic and *T*imely (in which it is associated a deadline). In creating your document, you will need to identify activities of high priority, particularly the ones being labelled as urgent and important. Put these at the top of the list. Less urgent and less important activities can be placed last. Furthermore, it is best that you attempt to do hard things first and easy things last. By doing so, you will give yourself sufficient time to identify and solve any unforeseen problems along the way. You can also categorise activities and put them under different headings, as illustrated below.
3. Updating your Word document (from point 2): Ideally, before you begin work each day, you can refer to the Word document that you have developed and to regularly update your to-do list. As soon as a piece of work comes in that requires your attention, you will need to look at your work schedule and re-prioritise if necessary. Do not be afraid to change your priorities or time deadline in relation to completing a specific task. Once you have finished a particular task, you can then delete it from the list.
4. Getting organised before you travel for a meeting. For example, you can create a travel folder that you can take with you. At the front of the folder, you can write down meeting name, date and location. You can print out (and put into the folder) any relevant documentations, e.g. boarding passes, meeting agendas and any printouts that you need to read before attending the meeting.

My Day-to-Day Task
Project 1: Project Priority Nano
(Project code ABCD123)

- *01-Dec, 2016 Finish your contribution (Section 1 of the bid proposal). Submit results to science leader, Dr. A.*
- *10-Dec, 2016. ISO document. Get approval from consortium members and edit as appropriate. Give ISO document draft to task leader, Mr. B.*
- *Etc.*

Project 2: Project Micro
(Project code EFGH456)

- *20-Dec, 2016. Data logging into communal database. Inform organiser, Mr. C, to check validity of logged data.*
- *27-Dec 2016. Work in nanolaboratory. Run experiment number 41 and write up observations/results in laboratory book. Have your book signed by Dr. D. at the end of the day.*
- *Etc.*

Others

- *28-Dec 2016. Meeting with line manager, Mrs. E. Report on activities for the past month. Discuss and agree to what you need to do next.*
- *29 Dec 2016. Transfer all files from memory stick to C:\drive.*
- *30 Dec 2016. Reimburse travel expenses from San Francisco conference.*
- *31 Dec 2016. Make travel folder for meeting next meeting in Paris on 8 January 2017. Print out relevant documents, e.g. boarding pass, hotel coupon and agenda. Place folder on your desk ready to take with you on the day of travel.*

… Etc.

In addition to having an organised structure to your day-to-day activities, you can also instil some changes in your attitude and behaviour, so that you can manage your time better. As such (Forsyth, 2013):

- Accept the fact that it might not be possible to do everything on your to-do list. If this is the case, then you will need to manage other people's expectations, e.g. by not overpromising to clients.
- Do not dwell on the past. Always concentrate on the present so that you can affect future outcomes.
- Do not be glued to your e-mails (or phones). Remember, with most of the e-mails that enter your inbox, you do not need to answer them straightaway. If possible, dedicate a set time during the day when you can read and reply to e-mails, e.g. possibly once in the morning, then another slot before you go home.

- Ensure that you create the right environment to successfully complete certain tasks. For example, if you need to work in a quiet environment to write a management report, then ensure that you can create this environment ahead of time. For example, you can opt to wear noise-cancelling headphones (or have earplugs), put a do not disturb sign on your door, block out your calendar (i.e. label the time as busy), not answer the phone (but ensuring that you have an answer phone to pick up any important messages), etc.
- Strive to collaborate with others in order to achieve certain goals/targets.

16.6 Summary

The goal of this chapter is to give you background information on project management and how to develop the right skill set in order to become a great project manager. Although it is true that you can easily learn the theory on project management from books (or a management course), it is not the same as putting things down into practice. In this chapter, I have concentrated on some of the biggest hurdles that you will face as a project manager. As such I have given you practical guidelines on how to:

- Deal with people effectively.
- Identify (anticipate) problems.
- Make informed decisions to solve problems.
- Manage your time.

References

Armstrong, M. (2013). *How to manage people*. London: Kogan Page.

Baguley, P. (2008). *Project management*. London: Hodder & Stoughton.

Barker, S., & Cole, R. (2009). *Brilliant project management: What the best project managers know, say and do*. Upper Saddle River, NJ: FT Press.

Cooke, H., & Tate, K. (2011). *The McGraw-Hill 36-hour course: Project management*. New York: McGraw-Hill.

Forsyth, P. (2013). *Successful time management: Creating success* (3rd ed.). London: Kogan Page Ltd.

Gido, J., Clements, J. P., & Baker, R. (2017). *Successful project management*. Boston, MA: Cengage Learning.

Gilman, H. (2013). *You can't fire everyone*. London: Portfolio Penguin.

MacDonald, J., & Tanner, S. (2012). *Successful people skills in a week*. London: Teach Yourself.

Manser, M. (2012). *Project management in a week*. London: Teach Yourself.

Mathis, B. (2014). *Prince2 for Beginners: Prince2 study guide for certification & project management*. San Bernardino, CA: CreateSpace Independent Publishing Platform.

Pries, K. H., & Quigley, J. M. (2012). *Total quality management for project management*. Boca Raton: Taylor & Francis.

Young, T. L. (2006). *Successful project management*. London: Kogan Page Ltd.

Chapter 17
Leaders and Teams

If I look back in history, to try and come up with names of those individuals who are great leaders, then my list will include Winston Churchill, Martin Luther King, Seretse Khama, Abraham Lincoln and Sukarno (first president of Indonesia) … to name a few. You may have your own list. They are your natural born leaders, all sharing certain traits, in common, namely charisma and the determination to change people's mindset (and as such change the world) forever.

Although you may not have the intention to lead a country, the ability for you to effectively lead is an important skill to have. Even if you are a junior member of staff, you may be asked to lead at some point of your career. For example, your boss may ask you to lead a small team of people in order to deliver a sub-task in order to support a big project. The need for you to be a leader will even be more crucial should you ever want to progress further with your career, e.g. if you want to become a university professor.

The goal of this chapter is to give you background knowledge on the topic of leadership, particularly what leadership means and the behaviour/skill set expected of you. This chapter will focus on a case study i.e. a particular situation in which you are likely to find yourself in: when you need to lead a small group of people. As such, I will discuss what teamwork means and present some practical tips on how to lead.

17.1 What Is Leadership?

For you to be a great leader, you will need to have the ability to (Gill, 2011):

- Identify problems, opportunities and challenges.
- Engage with your followers and build moral.
- Foresee hard times, being able to steer your followers in the right direction.
- Mediate between strategy and execution of that strategy.

R. Tantra, *A Survival Guide for Research Scientists*,
https://doi.org/10.1007/978-3-030-05435-9_17

- Identify core values or a set of guiding principles, in order to navigate your followers.

At this point, I would like to differentiate between being a leader versus a manager. Although one can argue that a good manager will need to have some level of leadership skill, he/she will be more focused more on the implementation of activities, to ensure the successful delivery of predefined goals/targets. As such, a manager will always want to play by the rules, always aiming to do the right thing (Price & Price, 2013). A leader on the other is less focused on the implementation but has the responsibility for surveying the overall environment, making sure that the correct strategic decisions are made to ensure that everyone is heading in the right direction.

17.2 How Do You Become a Good Leader?

The key to being a good leader lies in your ability to win the hearts and mind of your followers (as illustrated in Fig. 17.1).

In order to do this, past experts have identified the need for you to (Maxwell, 1999; O'Connor, 1994; Radcliffe, 2012):

- Be clear on your belief and value system.
- Have a clear vision for the future.
- Be clear on your strategy, specifically on how to make your vision a reality. As such you will need to ensure that you:
  - Choose the right people (to implement your vision). For example, you will want to employ good managers, who can reach targets.
  - Delegate and spend a considerable amount of time with those who will be responsible for executing your vision. When delegating, remember to clarify

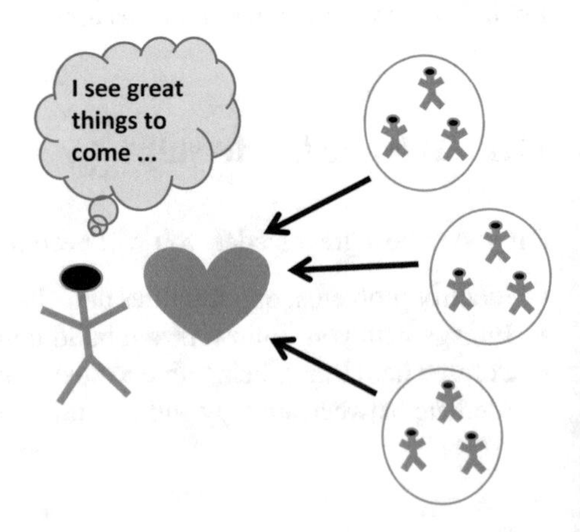

Fig. 17.1 Being a good leader means the ability for you to have a vision, on what the future holds and subsequently gaining support from your followers

goals and tasks, stating your expectations so that you can subsequently evaluate the outcome.
  - Foster a healthy environment, in which key people will be able to work together.
  - Identify and appreciate risks.
  - Prioritise and make informed decisions.

- Be consistent in your leadership style and how you treat people.
- Develop assurance, so that others will carry out your requests. Ultimately, your goal is to draw people to you. As such, you will need to:
  - Be confident.
  - Portray a sense of authority to be able to exercise power when needed, e.g. in order to align behaviours.
  - Value people, by treating everyone with respect (no matter what rank they hold). You can praise people (when they have done a good job) or give incentives for people to do better.
  - Communicate well, to deliver your message clearly and to infect others with your determination. Remember to listen attentively to what others have to say, especially in the context of your vision.
  - Be self-aware and be able to control yourself/your emotions. You will need to stay calm at all times and not express anger/upset, even if the situation is difficult, e.g. when people are shouting at you.
  - Dispense the kind of contagious energy that will help to induce: obedience, respect, loyalty and co-operation from followers.

In order to better highlight the main attributes on what it means to be a great leader, please refer to the illustration in Fig. 17.2.

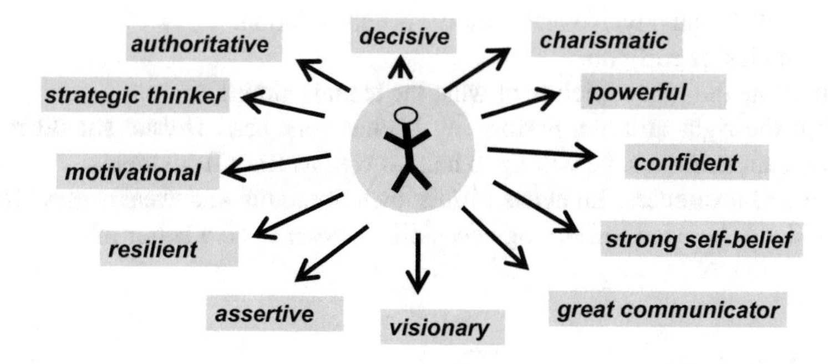

Fig. 17.2 Different attributes associated with being a great leader

17.3 Leading a Team

What is teamwork?

Teamwork exists when more than one person comes together, to work collaboratively to achieve goals/targets, e.g. to find a solution to a problem that is impossible to be achieved solely by an individual (Maxwell, 2001).

If you want to lead a team of people successfully, then you should strive to acquire some (if not all) of the essential personal attributes as identified in Fig. 17.2. According to an expert guru on the topic of teamwork (J.C. Maxwell), leading a team will mean the need for you to:

- Hire the right people for certain tasks. In order to do this, you will need to assess individual profiles, to identify those who can potentially act as sub-leaders, so they can ultimately share in the responsibility.
- Identify the *bad apples*. These are the troublemakers or those that seem to always let the team down. To begin with, you can always try to give them help and time to change. However, if things do not work out, you must be firm and remove them from the team.
- Communicate well with team members, e.g. what outcome you expect from your team members and to get their support or *buy-in*. You must ensure that you prevent any form of a communication breakdown within the team.
- Integrate people and establish team morale, so that people can work close to their full potential. In order to develop a sense comradeship, you must share victories as well as failures. You can also lead by example, such as sacrificing your own self comfort for the sake of the team.

As the leader for a team, it is important for you to know what behaviours you should expect from your team members. According to Maxwell, amongst other things, you should expect a team member to (Maxwell, 2001):

- Communicate well with other team members.
- Work collaboratively. No one should work in isolation.
- Flag any issues early on.
- Appreciate the bigger picture of what the team is aiming to do.
- Adopt the right attitude, performing his/her very best. He/she should respect other team members, be willing to help as well as learn from them.
- Have self-awareness, knowing his/her own strengths and weaknesses. He/she should also be open to learning new skills in order to do a better job.

17.4 Summary

Some people are natural born leaders but for many this is not the case. Yet, having a certain level of leadership skill is a necessary requirement if you want to progress with your career and hold a senior position. The aim of this chapter is to help you understand about what it takes to become a good leader. As such, I have given you background knowledge on the topic of leadership. In particular, I have discussed the kind of qualities/traits/behaviours that you should try to strive for, if you want to become a good leader. The chapter also gives practical tips on what to do, if you ever need to lead a team of people. As such, I have discussed the meaning of teamwork and how to become an effective team leader. Lastly, I have listed some of the qualities/traits/behaviour that you should expect from team members.

References

Gill, R. (2011). *Theory and practice of leadership*. London: Sage Publications.
Maxwell, J. C. (1999). *The 21 indispensable qualities of a leader: Becoming the person others will want to follow*. Grand Haven, MI: Thomas Nelson Publishing.
Maxwell, J. C. (2001). *The 17 indisputable laws of teamwork: Embrace them and empower your team*. Johannesburg: Struik Christian Books.
O'Connor, C. A. (1994). *Leadership in a week*. London: Hodder Education.
Price, A., & Price, D. (2013). *Introducing leadership: A practical guide*. London: Icon Books.
Radcliffe, S. (2012). *Leadership: Plain and simple*. Verlag: FT Press.

Part V
Looking for a Job?

Chapter 18
Curriculum Vitae (CV)

If you are looking for a job then you will definitely need to do a CV, as it will contain useful information about you, such as your education history, past achievements and work experiences. It is important to dedicate enough of time to write a winning CV, as this will ultimately be used by a hiring company to shortlist candidates potentially for an interview.

In this chapter, I will give you practical guidelines on how to develop two types of CV:

1. A (first) full CV: This CV is for personal use and should NOT be sent to the hiring organisation. It contains very detailed information and as such will have no page limit. In fact, you should put in as much detail as you wish into this CV. As you progress through your career, it is this CV that you will use to come back to, in order to update any new information.
2. A (second) targeted CV: This CV is formed from your (first) full CV and thus is suitable for sending out to the hiring organisation. It is written to target a particular job vacancy that you want to apply for. Hence, a targeted CV will be more focused and will have a page limit.

Nowadays, it is quite common for the hiring organisation to ask you to fill an application form instead of submitting a CV. As such, I will show you how you can update and use your CV, to fill in an application form. Whether you are sending out a CV or an application form, it is standard practice to submit an accompanying cover letter. This chapter teaches you how you can do this. Finally, I will present some of the common reasons as to why you can be rejected solely on the basis of your CV or application form. My intention here is for you to know what things to avoid, in order to improve your chances of getting an interview.

R. Tantra, *A Survival Guide for Research Scientists*,
https://doi.org/10.1007/978-3-030-05435-9_18

18.1 How to Develop Your (First) Full CV

As pointed out above, developing a (first) full CV is important, so that you can have all of the information right at your fingertips, for subsequent extraction of information to develop into a (second) targeted CV. Hence, when you see a job that you would like to apply, it will only take you a couple of hours (rather than days) before coming up with the final version, ready for submission.

When doing your (first) full CV, remember to add as much details as you want. You will never know what kind of information you will need to present to the hiring organisation. For example, in the education and employment section, you can include the day and month (rather than just the year) associated with start/leave dates. In the development of a full CV, remember to include (Mayer, 2008; Rogers, 2017):

1. *Personal details.* Here, you will need to state your name, address, telephone number, current occupation, e-mail address, nationality, etc.
2. *Education.* This should relate to your education achievements, usually from secondary school onwards. Detail: when/how long, where and subjects studied (including grades). Also include information on work experience, awards, scholarships, etc.
3. *Short courses* that you have taken in the past, detailing the type of course, grades achieved, where, when, for how long, etc.
4. *Employment history.* This will be the main highlight of your CV, as employers are always interested in what you can bring into the organisation from your work experience. Remember to include:
   (a) The dates (to/from) of employment.
   (b) Name of organisation and its location.
   (c) Position/level and salary.
   (d) Reason for leaving.
   (e) Achievements statements: You should list several achievements associated with every post. In trying to identify your main achievements, you will need to ask yourself the question of: what things have you been most proud of associated with the post? Be specific and give details. A good example of an achievement statement is: *Active in bid writing and subsequently secured around £3 million worth of funding over a 5-year period in protein array research.* This is better than just saying: *Experience in bid writing.*
5. *Professional qualifications.* This will include any affiliation or membership that you may have to a professional body.
6. *Profile.* If you have to sum up what you have to offer (in a few sentences), what would this be? You will need different personal profiles for the different jobs that you are thinking of applying. For example, if you want to work in the publishing world, then your profile will need to stress any editing or writing experience, rather than highlighting your scientific achievements.

7. *Other information* of relevance. You can list your peer-review publications, patents, grants awarded, hobbies, languages spoken, driving license, etc. Again, be as informative as possible. For example, in listing your peer-review publications, you can detail title, journal, year of publication, journal impact factor, citation index, etc.
8. *References*. You will need to list down five references, at least three from your most recent employer. Include details such as names, affiliation, position, work address, e-mail, telephone number and nature of relationship.

Please note that when developing your CV, there are a number of different types of format that you can adopt in order to present your information. Although there is not one strict format, I do recommend that you adopt a chronological CV format (as exemplified below). This type of CV lists the education and employment history in chronological order, e.g. with most recent job/post listed first. Employer usually prefers this type of format, as it tells your story about how you have progressed and grew over the years (Cortés & Mueller, 2007).

There is another type of CV structure that is quite popular and thus worth a short mention here, which is called a functional CV. This type of CV has less emphasis on who you have worked for in the past and what your position was at the time. Instead it highlights your skills and strengths; this type of format is suitable if you want to send out a CV purely on a speculative basis (Yate & Yate, 2003).

Name: …
e-mail: …
address/tel.: …

Personal profile:
A bioanalytical scientist with 5-year laboratory experience in HPLC, GC-MS … etc.

Employment history:
(2004–now) Company ABC, Senior Research Scientist
Technically led a team of 20 staff to deliver a large project (biopharmaceutical analysis using GC-MS) worth £500 k, over a span of 2 year time period.
… etc.
(2000–2004) Company DEF, Junior Research Scientist
3-Year laboratory research experience in HPLC and GC-MS on the analysis of biopharmaceutical products, delivering £200 k worth of contract for third parties.
… etc.

Education history:
(2001–2004) King's College, Chemistry, PhD. (pass), Chemical Research.
(1998–2001)University College, Chemistry with Biology, BSc. double honours (2:1).
… etc.

Professional qualifications:
(2004–onwards) Chartered Scientist.
(2003–onwards) Chartered Chemist, MRSC.
… etc.

18.2 Reading the Job Advert

Once you have developed your full CV, you will become more confident when it comes to searching for a job. Nowadays, most job adverts can be found online. Having said this, do not dismiss other sources of information, such as newspapers, magazines and trade journals. More importantly, remember that useful information can be gained through your own network, especially if you want to apply for a job that has not yet been advertised.

Once you have identified a job vacancy of interest, your next task is to read the requirements of the hiring company and then assess your own suitability for the post. You will need to carefully read the (Arnold & Barrett, 2017; Leigh, 2013):

- Company information: This can give you a clue about the culture, philosophy and ethos of the company. For example, if a company is into equal opportunity, then it tends to highlight this fact in the advert. Remember that the company information in the advert is always limited. As such, you will need to investigate further, by visiting its website or other sources, e.g. Glassdoor. From its website, you can read its mission statement, recent news/views section, etc. Be creative on how to search for information. For example, if you have personal contacts, you can talk to current/past employees to find out more about the company.
- Salary information: If salary details are not given, then it usually means that salary negotiations will only begin once a suitable person for the job has been identified. In addition, it is possible that the salary may not be an attractive feature of the post. If you want to have a rough idea of what is considered to be a reasonable salary for a particular post, then do your own research, e.g. finding out what the marketplace is currently paying for a similar position.
- Person specifications: This part of the advert is a vital read, as it describes the kind of person that the hiring organisation is looking for, in relation to qualifications, skill set, work experience, etc. The person specifications are usually divided into two parts: the essential and the desirable part. Remember that must fulfil ALL of the essential part; otherwise there is little point in applying. With regard to the desirable part, you will want to fulfil at least 80% of what is needed, remembering that these will be used by the hiring organisation to further filter candidates.
- Job specifications: This part of the advert is also important, as it details what you are expected to do in the job (should you get it). Unless extensive training is offered, the hiring organisation will usually want someone who has done similar roles before.

Once you have read the advert, you will have some ideas on what the hiring company wants and what it can offer in exchange. At this point, you will need to decide if you want to apply. In general, you should proceed if:

- There is a good match to what they want versus what you can offer.
- You do not have any issues with the job itself. There are various reasons as to why you may not want to pursue the job further, such as poor pay or temporary contract. Figure 18.1 exemplifies some of the factors that you may want to consider before applying.

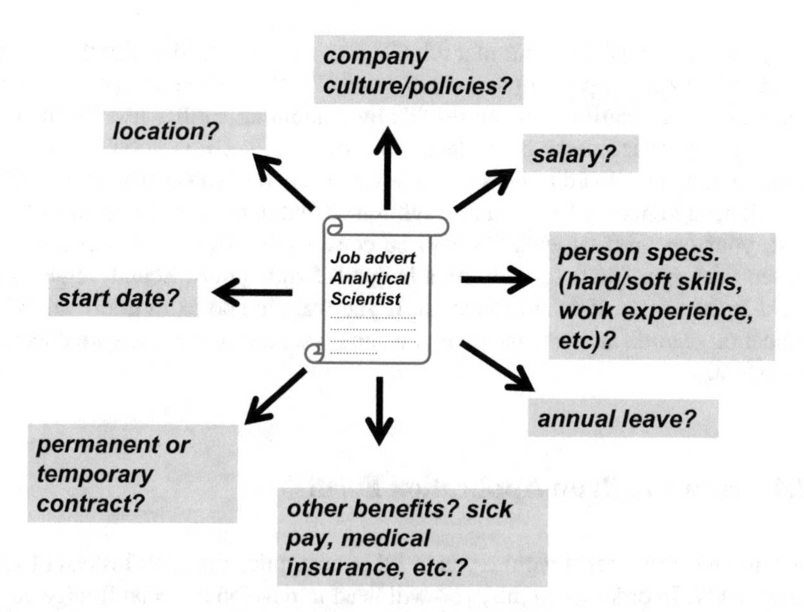

Fig. 18.1 Factors you may want to consider, to decide if you want to apply

If after reading the job advert you are still unclear, as to what you want to do, then do not hesitate to contact the hiring organisation directly with your questions. Adverts usually come with an e-mail address of a contact person. It may also be worthwhile dropping them an e-mail anyway to introduce yourself, prior to sending out your CV/application form.

18.3 How to Develop Your (Second) Targeted CV

Once you have decided that you will want to apply for a particular post, then you will need to develop, what I refer to as, your (second) targeted CV. This is the final, polished CV that you will send to the hiring organisation. In developing your targeted CV, you will need to read the job advert carefully and refer to your full CV at the same time. Your task now is to match what is needed from the job advert as to what you can offer in relation to, e.g., qualifications, skills, past experiences and achievements to support your case (Hobbs, 2013).

Using your computer, you will need to make a list of bullet points to highlight what you can offer (in relation to what is needed). I will refer to this list of bullet points, as *the matching list*. Once you have completed your *matching list*, you will need to highlight the entire text. Then cut each highlighted bullet point and paste it into the appropriate section of your (first) full CV. Ensure that any transferred text still remains highlighted, so that you are aware that this is the part of the CV that is most important to the hiring company. Once you have filtered all of the information from your *matching list* to your first full CV, you will then need to edit the information in order to develop your (second) targeted CV. You will need to edit for structure, content and clarity. You can do this by combining similar text together and removing any unnecessary or irrelevant information, such as hobbies. For more information on how to edit, please refer to Chap. 11. Remember that when editing, you will need to mainly focus on the highlighted text portion of the document. As a result, your (second) targeted CV will be considerably tighter, shorter and more focused than your (first) full CV. As a ballpark figure, your (second) targeted CV should be two sides of an A4 in length. If you want to add extra information, e.g. publication records and patents, then you should send this as a separate document attachment.

18.4 How to Fill an Application Form

Some job advertisements require you to fill in an application form instead of sending out a CV. In order to do this, you will need to develop a (second) targeted CV (as detailed before), based on the requirements of the advert as well as what is needed by the application form. Overall, you will need to prove that you can do the job successfully. Remember to edit your (second) targeted CV prior to transferring the appropriate text from this CV straight into the application form.

Although different employers will have their own forms, the requested information is often laid out in a logical/progressive order and constructed to reflect the job offer/requirements and ethos of the organisation. Typically, an application form can contain the following elements:

1. Personal data, e.g. name and address.
2. Educational background.
3. Employment history, along with the description of tasks associated with each post.
4. Question/answer section, to further assess your suitability: Remember that when answering these questions, you will need to refer to the person and job specifications (as set out in the advert). Typical questions can include:

   (a) What skills do you possess that will support your case?
   (b) Why do you think you are suitable for this position?

5. A list of references.
6. Professional qualifications.
7. Other relevant material: If you have additional information that can support your case, e.g. publication list, then you will often be asked to send this as a separate document attachment.

18.5 How to Write Cover Letters

A cover letter is a formal letter that is usually sent to accompany your CV or application form (Innes, 2009). As it is formal, the tone of the letter should reflect this. As such, your writing style needs to be clear and to the point but at the same time polite.

A suitable cover letter template is exemplified below:

[Your Name] …
[e-mail] …
[address/tel] …

[Recipient Name] …
[Company name] ….
[Address] …
Re: Job vacancy - research manager.
Dear Sir/Madam or Dear Mr./Mrs. …,
I have read your advertisement for research manager with great interest. Therefore, I am writing to outline the skills, knowledge and experience that I have gained for nearly 10 years as a project manager in the pharmaceutical industry … etc.
During my time at Pharmaceutical Company 1, I was responsible for … etc.
I believe that my skills, particularly with my previous experience in GLP/GMP, will make me an ideal candidate for the job … etc.
Thank you for your kind attention. I look forward to hearing from you.
Yours faithfully or Yours sincerely,
Dr. X.
Enc. CV.

In the body of text, you will need to cover key relevant points, such as stating the reason for writing and what documents you have attached with your cover letter. You can also briefly highlight your interest in the post and why you might be a suitable candidate. When doing this, remember to refer to the job and person specifications, to ensure that you closely match your skills/experiences to the requirements. Finally, remember that if you start the letter with *Dear Sir/Madam* you must sign it off as *Yours faithfully.* If you start with *Dear Mr./Ms./Mrs. X*, then you will need to end the letter with *Yours sincerely*.

18.6 Reasons for Rejection

There are a number of different reasons as to why you may get rejected on the basis of your CV or application form. Most likely, there will be some kind of mismatch between the job requirements and what you have to offer. As such, I need to iterate on the importance of ensuring that the information that you give to the hiring organisation matches well to the job and person requirements. However, other reasons can also include the fact that (McGee, 2014):

- You may be applying for the wrong organisation, in which you do not fit into the organisation's culture or ethos.
- The hiring organisation does not understand what you are trying to say, e.g. due to technical jargons or abbreviations. Remember that your CV can be read by someone who does not have any technical background, e.g. an HR manager or a consultant from a recruitment agency. Hence, make sure that your CV is understood by others outside of your scientific discipline.
- Your CV does not look professional, thus creating a poor first impression, e.g. poor layout, grammar/spelling mistakes throughout and some text being hard to read.
- Your writing is vague and lacks impact, such as when describing your past achievements. Remember to be specific (as previously explained).
- You have missed the deadline for submission.
- Your application form is not complete, with certain sections missing or not being filled in properly.
- There are far too many applicants, resulting in more competition. Ultimately, your application will be judged and ranked relative to others.

If you do end up with a rejection letter, then your confidence can take a knockback. My advice is to accept the rejection early on, to recover quickly from the disappointment and move on to the next job application. It may also be helpful to ask for feedback (so you can learn from your past mistakes).

18.7 Summary

In this chapter, I have given guidelines on how to develop a CV and fill in an application, should you need to respond to a job advert. Overall, the strategy proposed here involves the need for you to:

1. Develop a (first) full CV, detailing personal information, skill set, past education, work history, achievements, references, etc. This first CV is very detailed, to cover every possible aspect that you may be asked when responding to a job advert.
2. Read a job advert carefully, paying attention to both the job and person specifications and to ensure that you can satisfy these.

3. Develop a (second) targeted CV (based on the information from your (first) full CV). The aim of the second CV is to highlight how well you have matched your skills, work experience, etc. to the requirements of the post. It is this second CV that will be sent to the hiring organisation.

The chapter then continues to give you practical guidelines on how you can also use your (second) targeted CV, to successfully fill in an application form.

Whether you are sending a CV or an application form, it is likely that you will need to send a suitable cover letter. When writing the cover letter, remember to be polite and be direct, highlighting what you have on offer and why the job interests you.

Finally, I have discussed why a hiring organisation may reject you solely on the basis of your CV or application form. I hope that you will be able to learn from these and avoid such common mistakes, so that you can improve your chances of getting an interview. Good luck!

References

Arnold, C., & Barrett, J. (2017). *Taking charge of your career: The essential guide to finding the career that's right for you*. London: Bloomsbury Publishing.

Cortés, L., & Mueller, K. P. (2007). *How to write a résumé and get a job*. New York: Simon and Schuster.

Hobbs, R. (2013). *Inspire a hire: Successful job-hunting strategies for everyone*. Bloomington, IN: iUniverse.

Innes, J. (2009). *Brilliant cover letters: What you need to know to write a truly brilliant cover letter*. Harlow: Pearson Prentice Hall.

Leigh, J. (2013). *Successful CVs and job applications*. Oxford: Oxford University Press.

Mayer, D. (2008). *How to write & design a professional résumé to get the job: Insider secrets you need to know*. Ocala: Atlantic Pub. Group.

McGee, P. (2014). *How to write a CV that really works*. London: Hatchette UK.

Rogers, H. (2017). *Guide to writing a C.V. the easyway: Conducting a successful interview*. Brighton: Easyway Guides.

Yate, M. J., & Yate, M. J. (2003). *The ultimate CV book: Write and perfect CV and get that job*. London: Kogan Page.

Chapter 19
Interviews

If you are lucky to be offered an interview, then much congratulations in getting over the first hurdle! Give yourself a pat on the back in reaching so far. However, you must remember that getting an interview does not mean that you will secure the position as there is much work for you to do in its preparation.

My goal for this chapter is to guide you on how to prepare for your big day:

1. How to do background research before your big day.
2. What types of interviews you are likely to encounter.
3. How to anticipate the kind of questions that you are likely to be asked.
4. How to answer questions, particularly if you are given a competency-type interview.
5. How to tackle difficult questions.
6. General dos and don'ts on how to behave and conduct yourself, before and during an interview.
7. What you should do after an interview. In particular, I will be focussing on two aspects, i.e. how to negotiate your final salary (should you get an offer) and how to handle rejections.

19.1 Preparing for an Interview

Whenever I go for an interview, I always feel as though I am being examined, and observed under the magnifying lens by others. It feels very much as if I am having my PhD. oral examination again. As with any kind of oral examination, I cannot show up unprepared; this is not only ignorant but also extremely stupid. Thus, my first take-home message to you is this: prepare, spending a few days at least to ensure that you do a good job.

R. Tantra, *A Survival Guide for Research Scientists*,
https://doi.org/10.1007/978-3-030-05435-9_19

19.1.1 Doing Your Research

The first thing that you should do in order to prepare for an interview is to gather all of relevant information and put them in one folder. In this folder, you should have printouts of:

1. Your CV or the application form that was sent to the hiring organisation.
2. The job advert, highlighting the job and person specifications.
3. Company details, e.g. its culture, mission, branding and business model.

Undoubtedly, you will already have the information associated with points 1 and 2. With regard to point 3, you will definitely need to do more background research, as your knowledge of the company is likely to be confined to the information that was placed in the advert. Your next task therefore is to build a better picture of the hiring organisation in relation to the advertised position. You can get this information by visiting the company's website, talking to past/current employees, reading published articles about the company, etc. Ultimately, you will need to identify (Benisti, 2017):

- The products or services that they sell.
- Why they exist in the first place.
- Their customers' profile.
- Their mission statement.
- How do they do business. Their business model?
- Their core values. What are their employment policies like?
- Etc.

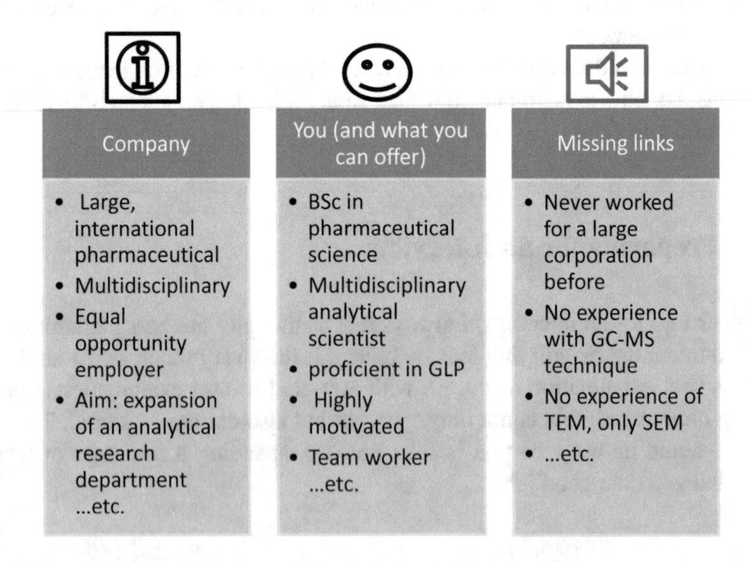

Fig. 19.1 An example of a tabulated information, summarising information on the company, what you can offer and your shortcomings

Fig. 19.2 Your success in a job interview will depend on how well you can match the requirements of the job as well as how well you fit into the culture of the hiring organisation

Once you have collected the information for points 1, 2 and 3 above, you will need to extract, condense and tabulate all the relevant information onto one side of an A4, as exemplified in Fig. 19.1. As shown, the tabulated information is divided into three separate columns detailing a summary (in bullet points) on the:

1. Company information.
2. Information about you and how you fit into the job requirements: When identifying the information about you and what you can offer, you must refer your CV/application form that you have sent and highlight the relevant qualifications, work experience, skill set, etc.
3. Any missing links, i.e. the specific requirements that the hiring company is looking for in an ideal candidate, but you currently cannot fulfil or satisfy. If you have a lot of missing links, then do not be too surprised if you do not get a job offer in the end; you are actually lucky to have been given an interview in the first place.

Just like a jigsaw puzzle, the hiring company will be focusing on how well you fit into the job, as well as into the culture of the hiring organisation (Vincent, 2013) (as depicted in Fig. 19.2). As such, it is vital that you are familiar with the information on the table. Make sure that you can elaborate further on the different points. At the interview, it might also help to visualise your tabulated information, to use it as a guide to constantly remind you on what is important to the hiring organisation. Remember not to ignore information under the missing links column and ensure that you can articulate any shortcomings that you may have. For example, if you are missing an essential work experience, you may want to remedy the situation by drawing attention on some similar tasks that you may have done in the past, so that you can at least discuss any shortcomings on a positive note.

19.1.2 Types of Interview

Before you show up for your interview, it is important to know what kind of an interview you will get. There are many different types of interviews, to include (Mayer, 2011):

1. One-to-one interview. As the name implies, you will be interviewed by one person.
2. Panel interview, in which you will be interviewed by more than one person: A panel interview can be more intense in that the questions can be fired one after another from different perspectives, by different people. For example, you may have a HR manager who will be interested on how well you fit into the organisation. A technical person on the other hand may be brought in to test your scientific knowledge. You might also be face to face with your prospective line manager, who will not only assess your competency for the job but will also scrutinise your attitude/character, e.g. to see if he/she can get along with you.
3. Group interview: This is when the hiring organisation tends to bring in a group of potential candidates together, so that they can be assessed simultaneously. If you are asked to come to this type of interview, it is likely that you will be divided into teams and given specific tasks to do, in which your behaviour will be monitored closely by the hiring team. This hiring team will want to see what role you naturally adapt to. For example, are you a natural leader or just a loyal follower? They will want to assess the skills that you may have. If you are in a group situation, remember that it is best to respect all those in your group and never adopt an aggressive tone to get your point across. Most organisations will want to hire assertive people rather than aggressive people.

In general, different organisations will have different preferences as to what type of interview to give in order to assess a candidate's competency for a job. Sometimes, you may need to attend more than one interview; it all depends on the job position and the hiring organisation. If you are given more than one interview, then it is likely that you will be invited for a distance-type interview first (e.g. through phone or video link), as a way to further shortlist candidates for the next round (Arnold, 2012). It is also possible that you will be invited to an assessment centre, where you can spend a couple of hours to a day there. If you do go to an assessment centre, then you will be subjected to a number of different interviews and often required to do an aptitude or a personality test questionnaire (Tolley & Wood, 2011).

At this point, I need to iterate my original take-home message: whatever type of interview you get, the hiring company will want to know how well you will be able to fit into the job and the organisation. As such, they will take into account various factors, namely your:

- Personality and behaviour traits.
- Set of soft and hard skills.
- Work experience.
- Education background.
- Motivation for the job on offer.

19.2 Questions

Let's face it: we do not have a crystal ball. You can never be fully prepared as to what kind of questions you will get in an interview. From my own experience, you can be asked anything (for any type of post), ranging from some common/expected ones (see below) to some bizarre questions, e.g. if you are given £5 million from this organisation, what would you do with it?

Whatever question the hiring organisation has for you, remember the following tips in order to help you formulate a good answer:

1. Always try to visualise your tabulated information, to:
   (a) Remind yourself about what is important to the hiring organisation.
   (b) Tailor your answer, so that it coincides with the company's vision and the job/person specifications.
2. Be an active listener and try to appreciate why the question was asked in the first place (Rizvi, 2008). For example, in the *£5 million pounds*-type question above, you may think that the interviewer is joking around and can dismiss it as a silly question. However, you will need to think hard as to why this question was asked and what it means to the interviewer. By asking the *£5 million pounds* question, it is likely that the interviewer will want to throw you at the deep end, to assess how you react under pressure. Hence, your best strategy here is to remain calm and answer as best as you can. A good answer for example, is to give a picture on how you would spend the money, considering the business model of the hiring organisation and their mission. Again, try your very best to link the answer to your tabulated information.
3. Answer the question as accurately as possible and remember to end on a positive note, especially if you have any shortcomings. Do not be pessimistic and sarcastic and always adopt a *can do* attitude (Hawley & Zemke, 2001).

19.2.1 How to Prepare for Your Answers in Advance

Although I have made the point that you can never accurately predict the kind of questions that will come up in an interview, it is always worthwhile to try and anticipate these questions in advance. Generally, there are two types of questions that you are likely to be asked: those that are tailored to the specific job/person requirements and those that are considered to be generic questions.

In order to predict what kind of questions are likely to come up, you must refer to the job/person specifications and pretend that you are the interviewer. By doing this, your mindset will become more in tune on the things that really matter to the hiring organisation. Try to come up with a list of 20 questions that you are likely to be asked and have some answers prepared in advance.

In addition to specific questions, you will also be asked some general questions too. Some common ones can include the following (Lees, 2012; Yate, 2017):

- So, tell me a little bit more about yourself. (Note: This is a common question that an interviewer often asks at the start of an interview, in order to break the ice. Remember that this is your 30-s commercial, in which you should highlight your unique selling point, of relevance to the post.)
- What are your strengths/weaknesses? (Note: This is to assess the level of your self-awareness. Remember to focus on those strengths that coincide with the job/person requirements. When talking about your weaknesses, make sure that these will not hinder you from doing your job properly, should you get hired.)
- Where do you see yourself in 5 years' time? (Note: This type of question aims to evaluate your ambition/vision and if these coincide with what the company can potentially provide.)
- Why do you want this job? (Note: This assesses your enthusiasm for the job and in joining the organisation.)
- What will you be bringing to the post and to the organisation? (Note: The interviewer here wants to know what you have to offer, to ensure that you skills/experience aligns with the job requirements.)
- Referring to your CV, I can see that you have done … *blah blah blah* … Can you tell me more about this? (Note: The interviewer wants to test if you have been truthful with your CV information. He/she will also want to assess the depth of your knowledge further. This is particularly the case with any technical based posts. For example, if you claim to be a scanning electron microscopy (SEM) expert, then the interviewer will want to make sure that you can talk about the technique in great details, e.g. how an SEM image is constructed.)
- Why are you thinking of leaving your current post? (Note: The aim here is to assess your professionalism and vision. Please do not say ANYTHING negative about your current job, boss, colleagues, etc. Give the impression that you are craving for future challenges.)

19.2.2 Competency-Type Questions

This type of interview is becoming more and more popular nowadays (Kessler, 2006). It aims to evaluate your competency, i.e. your ability to do the job. With this particular type of interview, the interviewer will ask you about your past experiences as evidence of your competency. The focus is on your past behaviour, which is an indication on how you will behave in the future, when confronting certain challenging situations. Hence, the hiring organisation will want to deduce if you:

1. Can do the job, i.e. do you have the necessary skill set/background experience to be able to do the job?
2. Want to do the job, i.e. do you have passion and drive to want to do the job in the first place?

You will know when you have been asked a competency-based question because you will need to describe past events, the challenges that you have faced and how you have managed to overcome them. Here are a few examples of some common competency-type questions:

- *Tell me a time when you have worked in a team and had to solve a difficult problem. What did you do?*
- *Describe a situation when you had failed a particular task miserably. What were the challenges? What did you do and what lessons were learnt?*
- *Etc.*

Here are some pointers as to what you can do to prepare for a competency-type interview:

1. Refer to your tabulated information and put yourself in the shoes of the hiring organisation, to appreciate what the interviewer is looking for in a person.
2. Choose five case studies (or experiences), in which you can elaborate during the interview that will highlight your competency for the job. For each case study, make sure that the achievement is of relevance to either the job or the person requirements, as advertised by the hiring organisation.
3. Practise articulating your case studies by adopting a *CAR* approach that stands for *C*hallenge *A*ction *R*esults (as illustrated in Fig. 19.3) (Roderick, 2012). You start off by describing the situation, detailing the challenges. You will need to say what you did, particularly how you overcame the challenges, before finally describing the outcome. Ideally, the result should be sufficiently detailed to portray you in a positive light. For example, if you have managed to secure funding as a result of your action, then you will need to specify how much and in what sort of timeframe. If for some reason you had chosen an example with a negative result, then you will need to say what went wrong, what lessons you have learnt and how you can do better in the future.

Fig. 19.3 The CAR (challenge, action, results) approach, a strategy on how to answer questions in a competency-type interview

19.2.3 How to Handle Difficult Questions

If you are struggling with a question, remember to pause and ask for clarification. This extra time will give you more time to think of an answer. However, if the question is obvious and you really do not know the answer, then you need to own up. It is better than waffling, trying to bluff your way out. This is especially the case if you are asked a difficult technical question.

Always strive to tell the truth. For example, if you do not have experience in something, then you must say so. However, try to end your answer in a positive light if possible (Arnold, 2012), as exemplified by the following:

Although I have no practical experience of TEM, I did an entire course during my undergraduate on this analytical technique and therefore have extensive theoretical knowledge. I found the course particularly interesting and thus am keen to put theory into practice.

19.2.4 Multiple-Choice Questions

If you know that you will be given multiple-choice questionnaires, find out what these are in advance. You will need to make sure that you are familiar with the type of questions that are likely to come up and ensure that you do a practice run beforehand. Sometimes the hiring organisation will want to give you an aptitude test, such as verbal reasoning, numerical reasoning and spatial reasoning. These types of tests will try to evaluate how your brain operates. You may also get a personality-type questionnaire. The aim of this is to delve deeper into your personality, to see if your personality is a good fit for the job (Parkinson, 2010). My advice with regards to this is to be as truthful as possible, as you cannot hide your personality once you start the job.

19.2.5 Your Questions to the Hiring Organisation

At the end of an interview, the interviewer will often ask if you have any questions that you would like to ask them. Your answer should always be a YES! As such, make sure that you have prepared some suitable questions to ask in advance. Do not ask too many, not more than three. Avoid asking weak questions, i.e. those of little relevance. Ideally, your questions should portray your genuine interest in the job and in the organisation (Fry, 2016). Some typical questions can include:

- What specific projects will I be working on?
- Who will I be directly reporting to?
- Etc.

19.3 Checklist of Dos and Don'ts: Before and During an Interview

19.3.1 Before an interview

1. Do your research thoroughly by:
   (a) Finding out more about the company and its mission.
   (b) Memorise your tabulated information (as previously discussed). Make sure that you can elaborate on this information in addition to any information on your CV or application form.
   (c) Be clear on what things are important to the hiring organisation.
2. Know who will be interviewing you. If possible, find out in advance the name and specific roles of the interviewers. For example, will you be dealing with a human resource person or a technical specialist? You can do a Google search to find out what their background and interests are.
3. Anticipate the kind of questions that you will get and prepare suitable answers in advance. If you will be getting a competency-based interview, remember to prepare a number of past case studies in order to support your case. If you are expected to do an aptitude test, make sure that you do a practice run beforehand.
4. Do a practice interview with a colleague or friend. If the hiring organisation has asked you to do a presentation, ensure that you have completed your slides in advance and practise several times beforehand.
5. Plan ahead, by making sure that (Duffy & Purcell, 2015):
   (a) You have the printouts of important documents to take with you on the day, e.g. printed version of your CV/application form, tabulated information, e-mail instructions from the hiring organisation and logistic information.
   (b) You do not show up late. Know in advance how you are going to get there. Plan to arrive 30 min earlier than the scheduled interview time. Remember to book a day off from work, so that you arrive in good time and have sufficient time to gather yourself before the interview.
   (c) You know what to wear and bring on the day. Plan to dress smartly and take care of your hygiene, e.g. breath and body odours. Take a visit to the hairdresser a week beforehand to smarten up. Also remember anything else you may need on the day, e.g. pen/paper for taking notes, umbrella (if raining) and fully charged mobile phone.
   (d) You calm your nerves. The day before your meeting, remember to put your interview folder away and just relax.

19.3.2 During an Interview

1. Create a good impression (Duffy & Purcell, 2015):
   (a) Switch off your mobile phone.
   (b) Do not eat chew gum, drinking or smoking.
   (c) Be friendly and smile but at the same maintain a professional outlook, e.g. do not joke around.

2. Be self-aware, especially when it comes to your body language (Human, 2017). You should aim to:
   (a) Strike a rapport with the interviewer/s as quickly as possible, e.g. by making eye contact.
   (b) Adopt the proper gestures and appear businesslike throughout. Do not make sudden movements or look at the floor or the wall. Sit upright and properly but do not be too rigid or take the defensive mode, e.g. folding your arms.
   (c) Relax. Find ways to control your nerves and prevent the buildup of adrenaline by adopting some suitable relaxation techniques, e.g. breathing deeply in and out and thinking positive thoughts. Refer to Chap. 1 for some tips on how you can do this.

3. Have a strategy in place so that you can answer the questions asked. Remember to:
   (a) Understand the intent behind the question before you answer.
   (b) Tailor your answers to the job/person requirements and the mission of the organisation.
   (c) Take a moment to think before answering, particularly if the question is difficult. Ask for clarification if needed. However, if you do not know the answer, be truthful and say so.
   (d) Show interest and enthusiasm for the job and the organisation.
   (e) Be confident but at the same time polite and courteous without acting cocky or arrogant. Talk clearly, being aware of your pace and tone.

4. Treat all panel members as equal, should you have a panel-type interview (Innes, 2015). Show your respect to everyone in the room, e.g. shake hands with all of the panel members rather than focusing on the one who you think is important, e.g. the one chairing the interview.

19.4 After the Interview

19.4.1 Negotiating the Offer

If a company is interested in you, then someone from the company will call you, to say that this is the case. Often, they will inform you what is on offer in relation to your starting salary. The topic of salary is a sensitive one but it is important that you accept an offer that you will be happy with, e.g. reflecting on the seniority of the position and the associated responsibilities. If you think that what you have been offered is far too low and does not reflect on your net worth, then you can negotiate, BUT only if you are in a position to do so.

You have two possible bargaining strategies to consider, in which the final figure can reflect on (Innes, 2015):

- The market value: In order to do this, you need to do your research and see what the market is currently offering for a similar position.
- How much others are willing to pay for your employment. If you have offers from other organisations, then you should inform the hiring organisation of this.

When bargaining, remember to be polite and diplomatic. If the salary offer is much lower than your expectations, then do not rush your decision. Ask the hiring company to give a day to think things through. It may be the case that you will need to meet them halfway, between what you initially asked for and what they are willing to offer. Remember, in deciding whether to accept the salary offer or not, you will need to also consider your position. For example, if you are currently out of work, then it's best to accept the offer for now, just so that you are back in the employment ladder. Furthermore, please consider the wider perspective, e.g. entire package, not just the salary (Innes, 2015). You will need to take into account what other benefits are being offered by the hiring organisation, e.g. childcare support, pension package and flexible working conditions. Also consider the kind of skills and experience that you will incur if you do decide to accept the offer, e.g. if the job will help to strengthen your market value and subsequently your future salary package. It is wise to accept a lower offer than what you originally plan for (or even currently earning) in order to secure a job that potentially has a lot of prospects.

19.4.2 Handling Rejections

If after all your hard work you end up with a rejection, then you will need to come to terms with this as soon as possible. Choose to move forward by (Walsh & Adams, 2007):

1. Accepting the rejection early on: Remember to:
   (a) Not take the rejection personally.
   (b) Be positive: Realise that perhaps it was not meant to be. It may be a blessing in disguise, as either the job or the employer may not have been right for you.
   (c) Get active, e.g. exercise, in order to get over your initial frustration.
   (d) Be kind to yourself and have self-compassion. Reward yourself. Remember that although you did not get a job offer, the fact of the matter is that your CV/application form has won you an interview. Consider this to be a partial success and this in itself is worth a pat on the back.
2. Staying focused in your quest to look for another job:
   (a) Learn from past experiences and mistakes, to prepare you better for the next round.
   (b) Do not put all of your eggs in one basket: Once you have finished with one interview, do not stop with your job search, as there is always a chance of a rejection. Think of alternative options all the time until you have secured a position.

19.5 Summary

When a hiring company is interested in your CV or application form, then you will usually get an interview, to further assess your suitability for the post. My goal in this chapter is to give you background information and practical tips, to help you improve your performance for your big day and help to secure the job that you want. In order to ensure interview success you must come fully prepared. As such, you will need to do the necessary background research, to find out more about the company, the things that are important to them, what their ideal candidate looks like, etc. It is worthwhile to anticipate several interview questions and prepare suitable answers in advance. If you have been invited for a competency-based interview, please ensure that you can articulate several relevant case studies from your past in order to support your application; remember to adopt the CAR principle when answering the questions.

This chapter also gives a checklist of do's and don'ts, on what you should do, before and during your interview. Once your interview is over, remember to:

- Ask questions to the hiring organisation, to show your interest in the job and the organisation.
- Consider any job offer carefully. Only bargain for your final salary if you are in a position to do so.
- Remain positive if you get a rejection. Choose to move forward and search for other opportunities as soon as possible.

References

Arnold, J. (2012). *Get that job with NLP: From application and cover letter, to interview and negotiation*. London: Teach Yourself.

Benisti, J. (2017). *How to succeed at job interviews*. Easyway Guides.

Duffy, M., & Purcell, T. (2015). *Survival guide to life*. Dublin: Spunout.ie.

Fry, R. W. (2016). *101 smart questions to ask on your interview* (4th ed.). Pompton Plains, NJ: Career Press.

Hawley, C. F., & Zemke, D. (2001). *Job-winning answers to the hardest interview questions*. New York, NY: MJF Books.

Human, H. (2017). *Body language magic*. Morrisville, NC: Lulu Press Inc..

Innes, J. (2015). *The interview book: How to prepare and perform at your best in any interview*. Harlow: Pearson.

Kessler, R. (2006). *Competency-based interviews*. Pompton Plains, NJ: Career Press.

Lees, J. (2012). *Job interviews: Top answers to tough questions*. Maidenhead: McGraw-Hill Professional.

Mayer, D. (2011). *Career essentials: The interview*. Valley Pub.

Parkinson, M. (2010). *How to master psychometric tests: Expert advice on test preparation with practice questions from leading test providers*. London: Kogan Page.

Rizvi, A. (2008). *Resumes and Interviews: The Art of Winning*. New Delhi: Tata Mc-Graw-Hill.

Roderick, C. (2012). *Interview answers*. Mumbai: Jaico Publishing House.

Tolley, H., & Wood, R. (2011). *How to succeed at an assessment centre: Essential preparation for psychometric tests group and role-play exercises panel interviews and presentations*. London: Kogan Page.

Vincent, T. (2013). *Nail that interview: Answer tough questions, make the best impression, and get the job*. London: Vermilion.

Walsh, R., & Adams, R. L. (2007). *The complete job search book for college students: A step-by-step guide to finding the right job*. Avon, MA: Adams Media.

Yate, M. J. (2017). *Great answers to tough interview questions*. London: Kogan Page.

Chapter 20
Redundancy

Being a Balinese at heart, my father often reminds me about life in general: what it is all about and how it works. An important lesson that I had learnt from him is that we all must accept the ups and downs of life. As such, like the food we eat, we will ultimately taste the different flavours of what life has to offer: sweet, salty, bitter, sour and metallic. Unfortunately, in today's competitive world, a job for life is no longer a reality and experiencing redundancy at least once in your lifetime often becomes the norm. Although some people may be (sweetly) compensated in a way that works for them, redundancy to most is not welcome news. In fact, it can leave the most complex intermingling of sour, metallic and bitter aftertaste.

This chapter is dedicated to the topic of redundancy, to enable you to cope should this unfortunate event ever happens to you. I will start the chapter off by discussing what redundancy means before giving you practical guidelines on what to do, in order to cope, i.e. to minimise the negative impact that redundancy may bring into your life. The chapter focuses on the three phases of redundancy, namely:

1. What to do before you exit your employing organisation and how to ensure that your redundancy process is a fair one. Please note that I will not be covering much about the legal requirements associated with this process, as it will ultimately be governed by your contract of employment and the legalities of the country that you are living in.
2. What to do after your exit from the employing organisation. This phase will be your biggest hurdle, as it is likely that you will be experiencing some kind of emotional turmoil. The discussion will thus focus on how to deal with your emotions and eventually overcome the grief of losing a job.
3. How to embrace the renewal phase. The focus here is on the practicalities, on how to be aware of the options that are available to you and how you can subsequently move on. Please note that I will not be covering topics that tells you how to look for your next job, e.g. how to write a CV, how to approach the marketplace and how to prepare for interviews, as these topics have been sufficiently discussed elsewhere in the book.

R. Tantra, *A Survival Guide for Research Scientists*,
https://doi.org/10.1007/978-3-030-05435-9_20

20.1 What Is Redundancy?

Redundancy (or layoff, in US terms) is when an employer dismisses you from employment, as your job or role has been made redundant and hence your services are no longer required (Corfield & Cushway, 2010). In fact, if you have ever gone through the process yourself, your employer will undoubtedly remind you that it is not the person but the job that is being made redundant and subsequently you are not being fired due to your incompetence in the job. There are a number of different reasons as to why a company will want to make people redundant. Quite often, the decision is governed by outside market forces, such as (Gilmore & Williams, 2013):

- The market having less demand for particular goods and services: As such, it is simply the case of taking into account changes in customer's demands and the need for the employing organisation to survive in a competitive market. The organisation can effectively shut down completely or implement some kind of a cost-cutting strategy. One option is to make some employees redundant to reflect the fact that part of the business has ceased to exist. Another option is to remove a senior member of staff (with a high salary) and to replace him/her with a more junior staff member (with a much lower salary).
- The fact that the employing organisation is merging with another company: This then forces the parent company to look at cost-cutting strategies, as the company will not need to have two separate departments to do the same thing. As a result, there is an intent to diminish part of the workforce to avoid duplication of activities.

20.2 Initial Phase of Redundancy: Being Practical

When you are being told by your employer that your job is being made redundant, you may feel as if it is the end of the world. As hard as it may be, you will need to deal with this initial shock quite quickly and put on a brave front. During this initial phase of the redundancy process, you will need to suppress any emotional turmoil that may be brewing inside of you. It is a time when you need to be logical (and less emotional) so that you can deal with practicalities before exiting the company. Just remind yourself that you will have plenty of time and opportunity to properly deal with your emotions after your exit.

Hence, the main objectives during the first phase is to:

1. Ensure that your redundancy process is a fair one and that you are getting the right compensation package.
2. Finalise any work that you need to deliver, and say your goodbyes before exiting the employment organisation with dignity and grace.

In order to achieve these objectives, you will need to (Gennard & Judge, 2005; Lees, 2010):

1. Understand the redundancy process. Your employer should inform you of:
   (a) The reason behind the redundancy.
   (b) How many jobs are likely to go.
   (c) How long the process will take.
   (d) Etc.
2. Be clear on what you are entitled to; this should be based on your contract of employment and what the legal requirements are.
3. Consider the possibility of voluntary redundancy, specifically if the company is willing to give you a better compensation package.
4. Consider alternative employment within the company. Remember that your employer may be able to use your skills in other areas of the business. Think twice before you decline such offers, as being redundant from a job will put you in a negative position in relation to your career prospects.
5. Be aware if you have been unfairly dismissed. Remember, it is not the person that is being made redundant, it is the job. If you think that you have been unfairly dismissed, then talk to a professional (e.g. union member, lawyer, citizen's advice bureau) to clarify your situation and seek legal advice.
6. Assess your financial position and figure out how long you can live on your redundancy package before you need another job.
7. Accept any help that your employer may offer, e.g. the possibility of being allocated free services from an external consultancy company that can help you with your CV, interviewing techniques, etc. This is especially useful if you have been out of the job market for a number of years. You will want to be in tune on how the job market works and what tools recruiters are currently using in order to assess a candidate's suitability for a particular post.
8. Complete all the necessary paperwork before you exit. Make a to-do list on what needs to be doing several weeks prior to your exit date, e.g. printing out your payslips/tax forms, having a written statement on how much you will be entitled to and having a character reference from human resources (HR).
9. Adhere to any rules imposed by your employer prior to your exit, e.g. clearing your desk and returning properties (such as laptops, staff card).

20.3 Dealing with Your Emotions

Once you exit the company, you can start to deal with your emotions, which may be the hardest part of the whole redundancy process. Redundancy can hit people in different ways, with some being able to cope more than others. It depends very much on how much you were attached to your last job. In circumstances where there is great deal of emotional trauma, remember not to shove your emotions under the carpet, in the hope that the feelings will go away. Instead, acknowledge the fact that you have gone through a major life event and that emotionally you may (Corfield & Cushway, 2010):

- Still be in a state of shock, confused and at loss. You may also still be in denial.
- Feel embarrassed, especially when you need to explain to family, friends and neighbours of your recent job loss.
- Feel sore, unwanted, rejected, as if you have failed.
- Feel angry and start to blame others for what happened.
- Feel anxious, depressed and have mood swings.
- Etc.

To some, facing a redundancy is like experiencing a bereavement of a loved one, possibly due to your loss of daily structure, identity, status/title, salary, etc. To some, the emotional upheaval becomes a daily battle. You may be experiencing the same type of negative emotions over and over again, as if you are in an *emotional washing cycle* (as illustrated in Fig. 20.1). If this happens to you, then you will need to be brave and deal with your emotions, which is a KEY first step.

I cannot say how long the *emotional washing cycle* will last, as it will depend on your personal circumstances, e.g. personality, outlook in life, how much you loved your last job and how soon you are able to secure another position. However, here are some tips that will help you to come to terms with your emotional baggage:

- Be aware of your emotions, i.e. identify what they are.
- Allow yourself time to grieve. Accept that it is okay to be emotional, as it is part of the grieving process. Do not feel guilty about taking time off in order to deal with your emotions and grief.
- Ensure that you are able to articulate your feelings. If your friends and family do not understand, then you must seek professional help, i.e. in the form a counsellor or therapist.
- If the redundancy has taken a toll on your physical and mental health, then talk to your medical practitioner, who can give you medications to help you cope with the symptoms of anxiety/depression (for example). Remember to look after yourself physically, as you will be emotionally vulnerable. Take care of your personal image and health (eat well, exercise daily, etc.). Please refer to Chap. 1 in this book for further information.

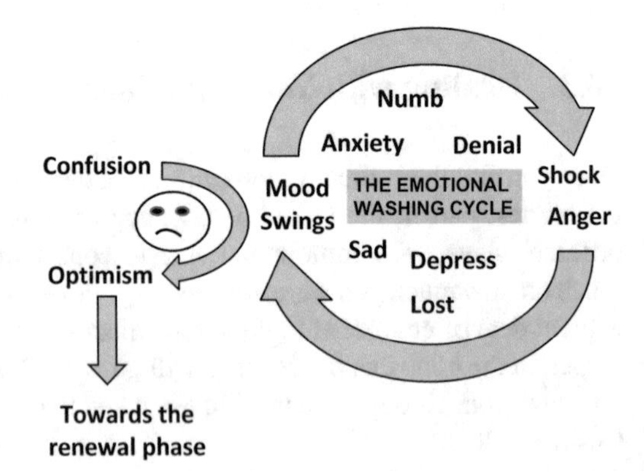

Fig. 20.1 Redundancy and the emotional washing cycle

- Ensure that you incorporate some kind of structure on a daily level, even though you may still be unemployed. For example, take the opportunity of doing some home improvements. Remember to incorporate some *me time* to enjoy life as well, e.g. going out for movie, eating in restaurants and playing uplifting music. Never punish yourself for being out of work.
- Acknowledge that your life has changed and you will need to adapt accordingly.
- Consider the possibility of having time out, a career break in order to improve physical, emotional and spiritual health.

The main goal for you during this phase is to deal with your emotions, to come to terms with what has happened. You will know when you are ready to move on, once your feelings of confusion are replaced with feelings of optimism. When this happens, you know you that you are on the right track and that you are ready to move on, i.e. heading towards the renewal phase (as indicated in Fig. 20.1).

20.4 Moving On

For you to move on in life, you must first appreciate the three different phases of the redundancy process, as illustrated in Fig. 20.2.

As previously mentioned, your toughest times will undoubtedly be during phase 1, when you are grappling with the emotional turmoil. As such, you will need to accept the fact that your recovery period from when you are told that your job will

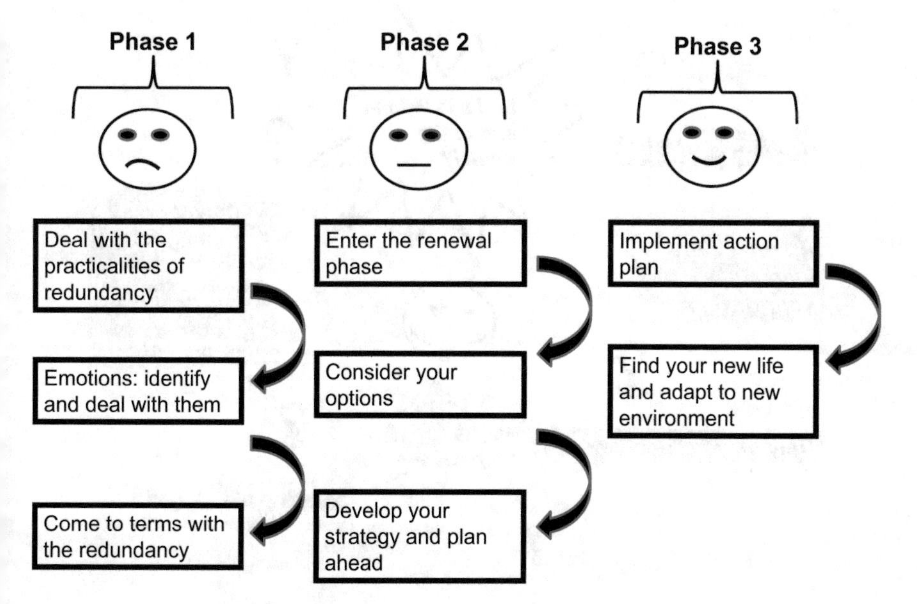

Fig. 20.2 The three phases of redundancy

be made redundant to a time in which you start to rebuild your life may take some time. A major milestone is reached, when you have noticed that your emotions have quietened down. You will feel that you have come to terms with what has happened and be more confident. The concept of moving on thus begins in phase 2, when you are entering the renewal phase. This is when you feel that you can rebuild your career, embrace the change and see the redundancy in a positive light. You will start focusing on what is working rather than what is broken, doing the things that you can do (rather than the things that you cannot). In this phase, you will assess your position, identify options and consider the opportunities (as illustrated in Fig. 20.3).

Remember that when you approach the marketplace, you will need to be realistic on what is required versus what you have to offer (see Chap. 2). Once you have made up your mind on what you want to do, you will begin to move into phase 3 (in Fig. 20.2), in which you will be implementing the necessary actions in order to carve out your new life, post-redundancy. You will need to remain positive, as it is possible that you may experience several knockbacks, which can take a toll on your self-confidence. Whenever possible, try to secure your next phase in life as soon as you can, so that you can regain your confidence. If you have trouble securing another post, then you will need to ensure that you do not have huge gaps in your CV, in which you are not doing much. Perhaps you will need to (Corfield & Cushway, 2010; Till & Till, 2003):

- Re-evaluate your strategy on how you search for a job (refer to Chap. 2). Are you getting repeated rejections because you are going for jobs that are not well

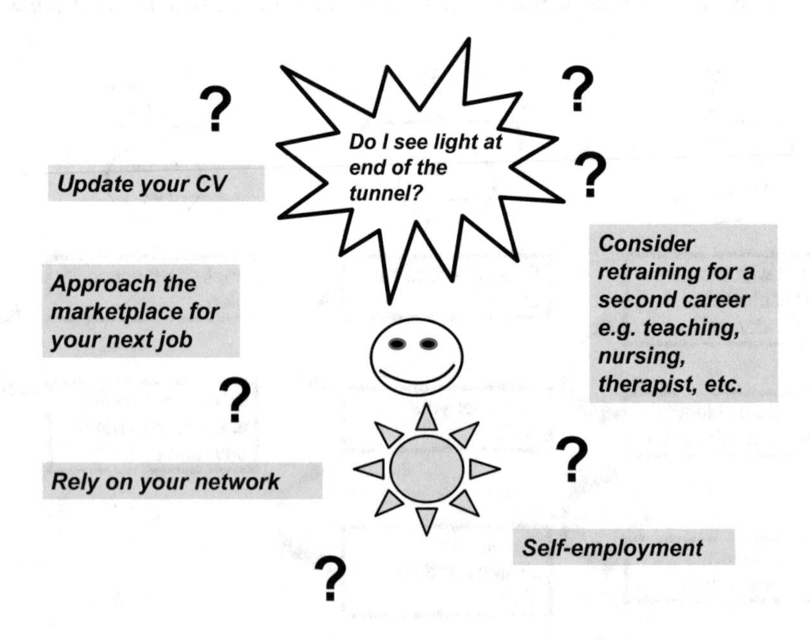

Fig. 20.3 Phase 2: Considering your options and identifying your strategy to move forward

matched for you? Is it possible to contact some of your old network, to give you advice on potential job vacancies, in which you may be suited for?
- Re-evaluate your career aspirations, to consider new career paths, with the strong possibility of re-training in a different area.
- Consider self-employment (to commercialise your skills and experience).
- Learn a new skill to add something to your CV, e.g. a new foreign language or IT skills.
- Take up voluntary work, potentially to diversify your skills in other areas or sectors.

20.5 Summary

To most people, redundancy is not welcome news and should it ever happen to you you must not underestimate on how this will impact your life. A huge hurdle in coming to terms with redundancy is the need to overcome any emotional turmoil that you may be experiencing. The goal of this chapter is to give you background information on the redundancy process and to give you practical tips on what to do, so you can cope better and move on. As such, I have given you guidelines on:

- How to deal with your emotions.
- What to do once your emotions have resided, to develop a more focused mindset and be practical (as you enter into your renewal phase).
- How to plan and implement future actions, so that you can find a new chapter in your life.

References

Corfield, R., & Cushway, B. (2010). *The redundancy survival guide: Assess your legal rights, explore career options and turn redundancy into opportunity*. London: Kogan Page.
Gennard, J., & Judge, G. (2005). *Employee relations*. London: Chartered Institute of Personnel and Development.
Gilmore, S., & Williams, S. (2013). *Human resource management*. Oxford: Oxford University Press.
Lees, J. (2010). *Career reboot: 24 tips for tough times*. Maidenhead: McGraw-Hill Professional.
Till, G., & Till, G. (2003). *The redundancy survivor's field guide: Use your redundancy to re-direct your career*. Oxford: How To Books.

Chapter 21
Self-Employment

When I became self-employed 2 years ago, I must admit that I did not know what I was doing. At that point in my life, I had just received redundancy from a science post that I did for 14 years. Although I had loved my job, I needed to move on and do something different and strived for a better quality of life. As such, I opted for self-employment. At the beginning, I looked forward to being on my own for a while, away from the regime of a daily routine. The idea that I can be my own boss (hurrah!), having the ability to dictate what I can do with my own time, attracted me to the prospects of being self-employed. Freedom is something I always cherish. However, as the reality slowly unfolds, I had learned (like many things in life) that there are always cons (as well as pros) associated with something that seemed (in the beginning) to be a lot greener on the other side of the fence.

The goal for this chapter is to give you sufficient basic knowledge about self-employment, so that you can decide, with your eyes wide open, if this is the right path for you. As such, the chapter starts off by highlighting the pros and cons for you to consider, before you decide to take the plunge. The chapter then aims to give you practical guidelines on how you can start your journey, once you decide to become self-employed. In particular, I will be discussing:

- Ways on how you can minimise the financial risks involved.
- How to develop a business plan.
- The key factors that governs business success.

Please note that I have no intention in giving you a course on business studies or bombard you with different theories on how to make your business grow. My aim here is to simply get you started towards the right path.

R. Tantra, *A Survival Guide for Research Scientists*,
https://doi.org/10.1007/978-3-030-05435-9_21

21.1 Why Become Self-Employed?

There are many reasons as to why you may want to head for self-employment, which may be due to either personal or professional reasons (or both). For example, you may (Burch, 2011):

- Not be happy with your current job, e.g. not being able to get along with your boss/colleagues, being bullied at work, not meeting your targets/goals, not fitting into the organisational culture, having lost faith in the appraisal system, being bored and unable to achieve a healthy work-life balance due to travel commitments.
- Want to have more freedom and be in control of your own destiny.
- Want to have a career break, to follow your passion or simply take time off to re-evaluate life.
- Believe that you have a product or service that is in demand and you want to explore its potential in the market.

As such, before you head towards self-employment, it is worthwhile spending some time to write down your own reasons, as illustrated below, which incidentally were my own reasons as to why I decided to become self-employed.

My Reasons for Being Self-Employed

1. When faced with the redundancy in 2016, this was one option that I was seriously considering. It was an opportunity for me to do something different.
2. I wanted to improve my quality of life and achieve a better work-life balance. I wanted to catch up on some quality time with my family/friends and to do the things that I enjoy, e.g. reading books, cooking and deepening my religious faith.
3. I wanted some me time, to explore and figure out what I want to do when I finally retire. As such, I opted to have a portfolio of different activities, such as:

   (a) Writing a book, i.e. this one.
   (b) Being a scientific consultant: From time to time I worked as an expert reviewer for the European Commission, to judge and evaluate bid proposals.
   (c) Taking my hobby as a soap maker to the next level, i.e. selling soaps to general public. I wanted to gain practical business experience.

21.2 Minimising the Risks Associated with Self-Employment

Let's get one thing straight: self-employment can be hard work and extremely risky. As such, you will need to appreciate what you have right now and what you will be losing if you do decide to give up the day job. For a start, you will be saying goodbye to the safe trappings associated with a regular monthly salary and other benefits that your employer may provide such as sick pay, holiday pay and pension contribution. When you go into self-employment, there is little guarantee that things will work out and you will need to appreciate the risks involved. Having said this, there are ways to minimise these risks (Jay, 2014; Johnson, 2011):

- Do NOT give up your day job straight away: If you can, start implementing your business idea whilst you are still in full-time employment, by doing things during the weekends or on your days off. Consider taking on a sabbatical leave. During this time, your goal is to develop your product/service and establish if it has sufficient demand in the marketplace.
- Do NOT go into business without knowing your product: Do your market research and ensure that you:
  - Understand the 4Ps of marketing, i.e. product, pricing, promotion and place. See below for more information.
  - Clearly identify the unique selling point of your product, i.e. how your offerings differ from competitors.
  - Release your product out into the market as part of a trial run to establish actual demand and get customer feedback, e.g. on how to improve your product.
- Do NOT go into business without being aware of the financial risks and how you can minimise them: As such, you will need to:
  - Prepare yourself when the business does not work out.
  - Save money and keep costs down at the start of your business. Do not spend money unnecessarily in hiring professionals such as a website designer. Learn to do things by yourself. If you do not have the knowledge, then you can learn by taking up a course or reading books. It might be worthwhile to contact your local council for possible support, e.g. free courses, grants and access to a business mentor.
  - Ensure that you do not have a cash flow problem. Do not spend all the money that goes into your business. Set some aside for things like national insurance, income tax and future product development.
  - Carefully consider before you decide to work with others: Do not rush into working with a partner in the early stages of your business, unless he/she is financially supporting your venture. Always opt for a trial period to see if you can work with that person.

   - Have the support of your partner (or family): Besides moral support, you may need to be financially reliant on your partner, to ensure that your family will have enough money to meet basic living needs.

- Adopt the right attitude: Be realistic and do not make assumptions about your product and on how well it will do in the marketplace.

21.3 Getting Started: Your Business Plan

Before you sell your product/service to the outside world, you must know what it is that you are selling and the market that are entering. By understanding this, you will be able to come up with a suitable business plan, which should be done before you decide to launch your business. There are four basic steps to help you develop a business plan, as detailed below.

21.3.1 First: Do Your Market Research

The purpose of a market research is to gather evidence, to show if there is sufficient demand for your product/service. Hence, when doing your research, you will need to answer several questions (Ali, 2002):

- Who are your competitors?
- How are your products different to what is already out there?
- Who are your customers? What are their profile, e.g. age range, location and shopping habits? Try to understand what makes a potential customer buy from you rather than going to one of your competitors. By understanding this, you will know how to entice your customers so that they will want to buy from you.
- Is there any way that you can further improve on your product offering?

There are two possible approaches that you can adopt in order to do market research. You can opt (Ali, 2002):

1. For a desk-based research: This involves accessing historical and published data, in which you can extract useful information such as size of your market or sector, identifying your competitors and knowing your potential suppliers.
2. Or carry out a field research: This is when you need to go out there, in order to get access to more accurate and up-to-date information. This may involve you giving people free samples and subsequently conducting a face-to-face interview with a sample population to get honest feedback and asking on how the product/service may be improved.

21.3.2 Second: Understand the 4Ps of Marketing and Doing a SWOT Analysis

Once you have collected your market research information, you will need to digest, analyse and present your data appropriately in order to come up with a marketing strategy. In order to do this, you must understand the 4Ps of the marketing mix (Wallace & Wallace, 2001):

1. *P*roduct. What is it that you are selling and to whom? Make sure that you have a clear product offering and that you can identify the unique selling point/added value associated with your product/service.
2. *P*romotion: What methods are you going to use in order to promote your product to your customers, e.g. via leaflet and website? How are you going to persuade your customers to buy? When you start promoting, remember to give evidence to highlight the added value/unique selling point, such as giving testimonies from satisfied/returning customers.
3. *P*lace. Where will you be selling your product/service in order to target your customers, e.g. online shopping and shops?
4. *P*rice. How much are you going to charge your customers?

Out of all the 4Ps, it is price that is most difficult to deal with. There is no golden rule associated with how to set your price, but you will need to (Timmons, Rhett, Weiss, Callister, & Timmons, 2013):

- Factor in your trading cost, e.g. raw material, insurance and salary. This is important, so you will not be selling at a loss.
- Have some ideas on your desired profit margin.
- Decide on your pricing strategy. If your main aim is to have high-volume sales, e.g. in which you are selling generic items, then you will need to set your price at the lower end of the price range. If you want to promote a high-quality/unique offering, then you can go for the higher end of the price range. You may also want to check out prices from your competitors and make sure that your price compares well with similar products/services already in the market.
- If you are really unsure about price, then identify a potential price range and opt for the higher end of the scale. You can always reduce your set price at a later stage.

A very useful tool to use in order to analyse/display your findings from market research is SWOT (an acronym for Strength, Weaknesses, Opportunities and Threats). The purpose of such an analysis is to identify (Hall, 2003):

1. The *S*trengths of your business, e.g. the added value of your product offering.
2. Any emerging *O*pportunities, in order to exploit what you have.
3. Any *W*eaknesses, so that you can find ways to minimise them.
4. Any *T*hreats to your business: As such, you will need to come up with appropriate contingency plans in order to buffer any negative impact.

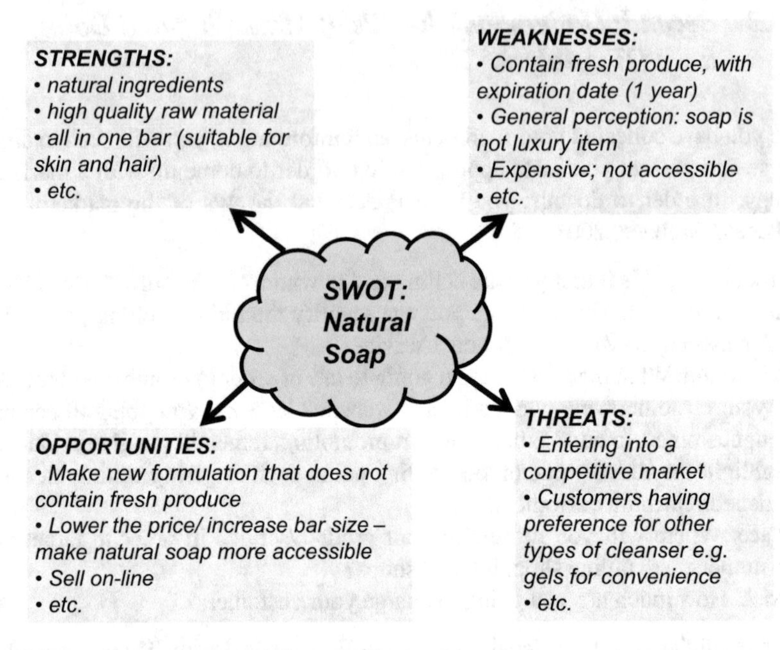

Fig. 21.1 An illustration of a SWOT analysis before I launched my own soap business

The SWOT analysis will enable you to clarify your marketing position, thus allowing you to establish an appropriate strategy in order to survive in the market. Remember to discuss your findings with a friend or (better still) a business mentor, so that you are able to see the bigger picture. Figure 21.1 illustrates an example of a SWOT analysis, which incidentally is one that I did before I decided to launch my soaps into the marketplace.

After conducting your market research, you should be in a position to establish if there is sufficient evidence to start a business. If it is not the case, i.e. there is little demand for your product/service, then you have the option to either amend/update your original idea significantly or abandon your initial idea all together and start all over again.

21.3.3 Third: Choose a Suitable Business Structure

You will need to decide what type of business structure you will want to adopt in order to run your business. This will be governed by your personal circumstances, the product/service on offer and how you want to conduct your business, e.g. do you want to have another person to help you run it? There are different types of business structures, with some common ones being (Fontana, 2009):

1. A sole trader
2. A partnership
3. A limited liability company

Whatever structure you do decide on, you must understand the legal and tax implications associated, such as the need to declare your business to the tax man. If you are unclear as to what you need to do, then you must seek the advice of a professional, e.g. solicitor, accountant or financial advisor.

Out of the three types of structure (above), the simplest by far is to become a sole trader, in which you will be working on your own (Fontana, 2009). In the UK, becoming a sole trader is easy to set up, as you simply need to inform the tax man and submit a tax return at the end of each tax year. If you cannot do this yourself, then you can hire an accountant to do it on your behalf. Being a sole trader will also mean that you avoid limitations such as extra formalities often associated with the other business structures. Hence, opting to become a sole trader is perhaps the best path to take, if you are still unsure about your business. Nonetheless, you should be wary of the limitations associated with being a sole trader, namely that (Reuvid, 2011):

- Business can be slow, as there is only one of you. You will be dealing everything by yourself, which can be time consuming and costly.
- You can feel isolated/alone, as you will not have anyone to talk to about issues.
- There is no distinction between you and your business. Should your business fold, then your creditors have every right to dissolve not only your business but your other assets as well.

Although being a sole trader is simple to set up, you will still need to acquire certain basic skills in order to manage your business, such as basic bookkeeping skills, so that you can note down all financial transactions that are made into/out of the business.

21.3.4 Fourth: Write the Actual Business Plan

This is when your ideas and market research findings are consolidated and put on to paper. The business plan will help you to develop a strategy on how you are going to carry out your business. The aim of the plan is to enable you to (Duncan, 2013; Reuvid, 2011):

- Establish that you have a valid business idea.
- Be more confident, so that you can move forward.
- Make mistakes on paper (rather than in real life).
- Identify how much money you will need in order to move the business forward and to justify the case. This is essential if you are seeking investors/collaborators.

You can put different types of information into a business plan. Ultimately, the content will depend on your business and who your readers will be (Hatten, 2012). If your reader is a potential investor, then you will need to tailor your business plan to accommodate his/her interest, such as presenting evidence that promises rapid future business growth. If your reader is a banker, then he/she will need assurance that your business will not fail. As such, you may need to portray a steady business growth, with risks having been fully identified and concrete plans on how to minimise these. Whoever ends up reading your business plan, make sure that you identify any threats/weaknesses and that these have been sufficiently addressed.

Although there is no set format with a business plan, it is usual to incorporate (Barrow, 2007; Pinson, 2008):

1. A front page, e.g. cover, title page and table of contents.
2. The business concept: This is when you need to describe the product/service offering and a description of the market that you are entering. You will need to highlight how your offering differs from that of your competitors. If you have a business that is already running, then you will need to delve into the history/background and update the current position of the company. Overall, you will need to give evidence that there is indeed a demand for your product/service.
3. The proposed marketing channels: You can present your marketing strategy based on the 4Ps and the SWOT analysis (as mentioned above). Remember to address any identified threats and weaknesses from such analyses.
4. Details on operations: Here, you will need to detail the day-to-day running on how goods/services are produced/delivered.
5. Disclosure of any patents or details of propriety position/issues.
6. Financial projections, which essentially are the forecasting of profit and loss, and cash flow (usually for next 3 years of the business).
7. Premises and location description.
8. Details on key staff, e.g. management team, their background and how they contribute.
9. Details on future requirements, for example if you need any money (and for what, e.g. staff recruitment, new capital equipment and new premises).
10. Executive summary or abstract, which is the overall summary of your plan and is usually placed at the beginning of the report. It is of particular interest to potential investors as they will often read this section first, in order to establish if the idea grabs their attention. If they are not interested in this part, then it is likely that they will not bother reading the rest of the document. Ideally, the executive summary should be one or two pages in length.

Remember that the market is never a static place. As such, you will need to be sensitive to any changes in the market demand, e.g. changes in customer behaviours due to recent trends. As such, treat your business plan as a living document. This means that you should update it regularly, particularly in light of any new information, such as any changes from the SWOT analysis. When updating and reassessing the business plan, you will need to come up with suitable action points to implement, so that you can remedy any changes in market demands. Overall, it is your

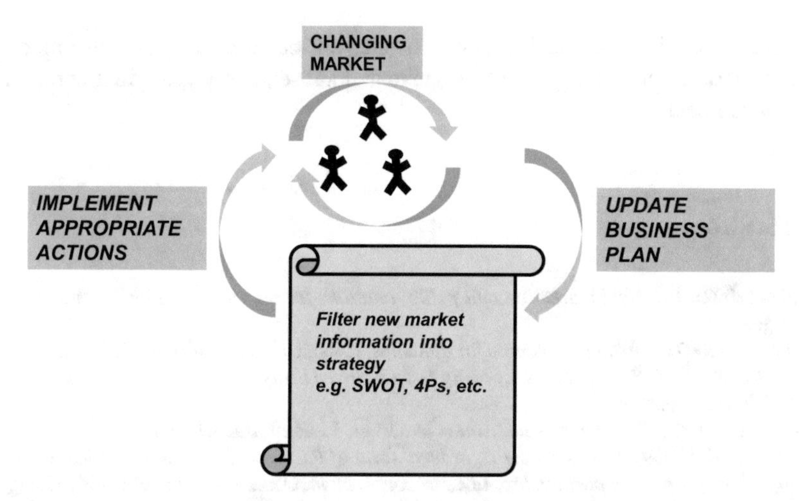

Fig. 21.2 Updating your business plan regularly is important, so that you can respond to any changes in the market conditions

ability to read and be in tune with the market, which will ultimately determine your business success (as depicted in Fig. 21.2).

21.4 Summary

This chapter gives you background knowledge on the topic of self-employment and what you will need to do in order to get things started. I have highlighted the pros/cons of being self-employed, so that you are aware of the things that you will need to consider before taking the plunge.

Some practical guidelines on how to begin your journey have been presented. In particular, I have discussed:

- Ways on how to minimise the risks associated with self-employment.
- Ways on how to conduct market research, in order to collect hard evidence to show that there is sufficient demand for your product/service, to ultimately help you decide if you want to launch.
- The need to choose a business structure that is right for you and ensure that you understand the tax and legal implications associated.
- How to do a business plan (even if you are not looking for investors).

Although there are many factors that can affect the success of a business, it is ultimately your ability to be in tune with the market that is so important. In order to achieve this, you will need to:

- Treat your business plan as a living document and update it regularly.
- Be sensitive to market changes, such as changes in customer needs.

- Adopt a flexible approach on how to run your business, so you can implement appropriate actions, in order to survive and subsequently grow in a competitive environment.

References

Ali, M. (2002). *Practical marketing and public relations for the small business*. London: Kogan Page.
Barrow, C. (2007). *Starting a business for dummies*. Hoboken, NJ: John Wiley & Sons.
Burch, G. (2011). *Self-made me: Why being self-employed beats everyday employment every time*. Chichester: Capstone.
Duncan, K. (2013). *Start your own business in a week*. London: Hachette UK.
Fontana, P. K. (2009). *Choosing the right legal form of business: The complete guide to becoming a sole proprietor, partnership, LLC, or corporation*. Ocala: Atlantic Publishing Company.
Hall, R. E. (2003). *Starting a small business: A step-by-step guide to help you plan and start a small business*. Haverford, PA: Infinity Publishing.com.
Hatten, T. S. (2012). *Small business management: Entrepreneurship and beyond*. Mason, OH: South-Western Cengage Learning.
Jay, L. (2014). *Self-employment--the secret to success: Essential tips for business start-ups: The beginner's guide to setting up and managing a small business*. L&J Business Solutions.
Johnson, L. (2011). *Start it up: Why running your own business is easier than you think*. London: Portfolio.
Pinson, L. (2008). *Anatomy of a business plan: The step-by-step guide to building your business and securing your company's future*. Tustin, CA: Out of Your Mind and Into the Marketplace.
Reuvid, J. (2011). *Start up and run your own business: The essential guide to planning, funding and growing your new enterprise*. Philadelphia: Kogan Page Ltd.
Timmons, M. B., Rhett, L., Weiss, J. R., Callister, D. P. L., & Timmons, J. E. (2013). *The entrepreneurial engineer: How to create value from ideas*. Cambridge: Cambridge University Press.
Wallace, D., & Wallace, S. (2001). *GCSE business studies*. Oxford: Heinemann.

Index

R. Tantra, *A Survival Guide for Research Scientists*,
https://doi.org/10.1007/978-3-030-05435-9

W

Printed in the United States
By Bookmasters